# QUARKS AND HADRONIC STRUCTURE

## INTERNATIONAL PHYSICS WORKSHOP SERIES

Series Editor: **A. Zichichi**
*European Physical Society*
*Geneva, Switzerland*

---

*Volume 1:* **QUARKS AND HADRONIC STRUCTURE**
**Edited by G. Morpurgo**

# QUARKS AND HADRONIC STRUCTURE

Edited by

G. Morpurgo

*Istituto di Fisica dell'Università*
*Genova, Italy*

PLENUM PRESS · NEW YORK AND LONDON

Library of Congress Cataloging in Publication Data

Main entry under title:

Quarks and hadronic structure.

(International physics workshop series; v. 1)
Includes index.
1. Quarks—Congresses. 2. Hadrons—Congresses. I. Morpurgo, Giacomo, 1927-
II. Ettore Majorana Centre for Scientific Culture.
QC793.5.Q252037 539.7'216 76-47490
ISBN 978-1-4684-0930-7 ISBN 978-1-4684-0928-4 (eBook)
DOI 10.1007/978-1-4684-0928-4

Proceedings of the International Workshop of the Ettore Majorana Center
for Scientific Culture held in Erice, Italy, September, 1975


Softcover reprint of the hardcover 1st edition 1977
A Division of Plenum Publishing Corporation
227 West 17th Street, New York, N.Y. 10011

## Foreword

One of the activities of the Ettore Majorana Centre for Scientific Culture is the international advanced study courses on scientific topics which are of particular relevance today. The Centre is located in Erice, a mountain town in the province of Trapani in Sicily. At present over seventy Schools of the Centre are active, holding annual or biennial courses, so that about forty courses are organized each year. To date some twenty thousand participants have attended the courses of the various Schools of the Centre.

The International Physics Workshop Series has been established to make the contents of the Workshops of great topical interest available to those who were unable to attend them. The courses are conducted on an advanced, post-doctoral level. This volume - the proceedings of the session on "Quarks and Hadronic Structure" - is the first of the Series.

In September 1975, thirty-three physicists from twenty-one laboratories in nine countries met in Erice to attend the Workshop. The countries represented were: Austria, France, Germany, India, Italy, Poland, Switzerland, the United Kingdom, and the United States of America. The purpose of this Workshop was to bring together a group of theorists working on various aspects of the quark structure of hadrons to discuss and critically evaluate the present situation.

Professor Morpurgo was given the direction of the Workshop. I would like to take this opportunity to thank him most warmly for having accepted this responsibility and for the success of the Workshop.

A. Zichichi

# Preface

A glance at the table of contents and the names of the authors will make it clear what this volume is about. There is no need for further illustration -- all I could do is repeat the substance of my introductory lecture, which was intended precisely to provide a sort of annotated index.

The task of organizing the Workshop and editing this volume was made easy by the cooperation of the lecturers. I am grateful for this to all of them. In particular, I thank warmly Professors R. H. Dalitz, H. Joos, and G. Moorhouse -- members of the organizing committee of the Workshop -- for their invaluable help in organizing the session and the details of its program.

I am also deeply indebted to Professor A. Zichichi, Director of the Ettore Majorana Centre for Scientific Culture, for the essential support given during all the stages of the Workshop, from the very beginning to the editing of these proceedings.

G. Morpurgo

Genoa, January 1977

# Contents

# OPEN PROBLEMS IN THE QUARK MODEL

G. Morpurgo
Istituto di Fisica dell'Università - Genova
Istituto Nazionale di Fisica Nucleare - Sezione di Genova

## 1. Introduction

After ten years from the formulation of the realistic quark model there is little doubt that the model has provided a considerable insight in the hadron physics; but it is equally sure that a large number of conceptual and practical questions remains open.

Probably most of these questions might be answered rather easily if we could meet the free quark and measure its properties; its mass, its charge, its magnetic moment etc. But, because this has not been so far the case (and, maybe, it will never be the case) we continue to be confronted with a strange and difficult situation, in some respects a paradoxical one: quarks are not seen but many things go as if they existed.

Indeed there is now little doubt that, unless the free quark is discovered, a satisfactory way out from this situation will require some radically new idea; an idea which, of course, will have to take into account the new spectroscopy being developed in these days after the discovery of the J (3.1) and of the $\psi'(3.7)$ about which we shall hear more in this meeting. Let me remark, incidentally, that this spectroscopy appears to show already, as a subproduct, the usefulness of the non relativistic concepts; also if we do not still know if the objects which are bound are charmed quarks, coloured quarks or perhaps heavy leptons interacting through some medium strong interaction [1].

Meanwhile it seems appropriate to clarify as much as possible the present status of the quark model so as to recognize clearly its successes, its limitations and its difficulties. This has been the motivation for this meeting; in these introductory remarks, I confine myself to list a number of questions which are going to be dealt in the talks and discussions which follow. I may take the opportunity to express my personal point of view on these questions but the main purpose of these short remarks will be to give a logical order to the set of questions which are going to be debated. The order by which I shall proceed will be from the more clear or simple questions to those which are less understood. Let me start with the classification of the states in the quark model.

## 2. Classification of the states

a) Mesonic. Here I refer to the talk by A.P. French [2] at the recent EPS Conference in Palermo. It is certainly very pleasant that in addition to the $3^-$ g(1680) meson a new h(2020) meson with spin 4 and parity + has been established independently by two groups, corresponding to an excitation of the L = 3 family. It is however unpleasant that the problems with the $1^+$, $0^+$ family all remain open. Only an optimist can, at the moment, see clear evidence for these mesons. One should continue to keep this point in mind.

b) Baryonic. I am not in the best position for analyzing the successes or the failures in the classification of the baryonic states in the quark model. The reason for this is that my attitude with respect to the assignement of resonances in $\pi N$ scattering based on phase shift analysis has been, since long time[3], somewhat skeptical so that I have not kept much track of the evolution of the situation. But Professors Dalitz and Moorhouse will illustrate, I am sure, the situation in this field.

## 3. Electromagnetic properties

Here the main static properties predicted by the model are: 1) the ratio -3/2 between the magnetic moments of proton and neutron. The 4% difference between this number and the experimental value - as well as the much larger percentage difference in the M1 $\Delta^{++} \to P\gamma$ matrix element - can probably be attributed to one or more of three reasons: a) $SU_6$ admixtures, b) use of relativistic corrections, c) exchange terms in the current. Similar problems, essentially unsolved after decades, exist in nuclear physics[4]. The other ratios between the magnetic moments of the hyperons, as well as the value of $\mu_{\Sigma^\circ\Lambda^\circ}$, which begins to be known, are consistent with, but add little to the comparison with the quark model due to their present level of precision. I am sure however that soon we shall have new important and precise data to discuss. 2) The essentially zero charge radius of the neutron; this is also easily predicted by the naive quark model.

Passing now to the radiative transitions let us comment separately on the mesons and on the baryons.

As far as the radiative transitions of the mesons are concerned, the $\omega \to \pi^0\gamma$ decay has been an hystorical success of the non relativistic quark model[5]. The calculation is free of ambiguities and a rate in good agreement with the facts is obtained treating this decay as a M1 decay, using the value of the magnetic moment of the quark obtained from the magnetic moment of the proton and taking the same space wave function for the $\omega$ and $\pi$.

I will come back to this example in my next lecture in connection with the discussion of how to construct a covariant quark model; but here I would like to insist on the importance of careful measurements of the other radiative decays of mesons.

I have in mind, specifically, the $\Phi \rightarrow \eta\gamma$ , $\rho \rightarrow \pi\gamma$ and $K^* \rightarrow K\gamma$ decays. All of them have been measured: the first is in acceptable agreement with the quark model, but aside from the large discrepancy existing between two different experiments[6], the satisfaction is somewhat softened by the possibility of playing with the $\eta$ , $\eta'$ linear or quadratic mixing angle.

No problem of this kind exists for the $\rho \rightarrow \pi\gamma$ and $K^* \rightarrow K\gamma$ decays; the relations of their matrix element to that for the $\omega \rightarrow \pi\gamma$ decay only depends on the quark model and, for the $K^*$ decay on the magnetic moment of the $\lambda$ quark. The present data[7] on these matrix elements are a factor $\sim 1.7$ (in both cases) smaller than the prediction. At the moment I do not attribute too much weight to this fact because (especially after the experience of the $\eta$ meson) we are well aware of the difficulties in separating the nuclear coherent from the Primakoff production. But if the above rates should continue to be a factor three off from the predicted ones in future experiments, this should become a point of some concern.

Let us now say very few words on the radiative transitions in baryons; here the situation is much more complicated and professor Moorhouse will illustrate all its complexities. In addition to the Dalitz-Sutherland[8)] transition already mentioned, I confine to note that the original simple selection rules predicted by the quark model (the Becchi-Morpurgo[9)] and the Moorhouse[10)] selection rules) hold well; the problems analized at present continue the more complex kind of quantitative treatment exemplified by Copley, Karl and Obryk[11)].

## 4. Semileptonic decays

I will insist a little on this topic because, in my opinion, it serves to clarify rather well many conceptual problems and in particular the distinction between current quarks and constituent quarks. In a non relativistic quark model description, the Cabibbo current, governing the semileptonic decays, arises in a very natural manner. Indeed, non relativistically, only Fermi and Gamow Teller transition operators are possible; so that, excluding induced terms, the most general charged current x current interaction containing a $\Delta S = 0$ and a $\Delta S = 1$ term has necessarily four and only four parameters (G, $\theta$ , a and b):

$$\frac{G}{\sqrt{2}} \left\{ J_0^{lept} \left[ \cos\theta \, (p^+ n) + \sin\theta \, (p^+ \lambda) \right] + \right.$$

$$\left. + \underset{\sim}{J}^{lept} \left[ a \cos\theta \, (p^+ \underset{\sim}{\sigma} n) + b \sin\theta \, (p^+ \underset{\sim}{\sigma} \lambda) \right] \right\} \qquad (1)$$

Barring induced terms - which are not in fact entirely negligible but tend to zero in the zero 4-momentum transfer limit - the above eq. (1) is simply the most general interaction one can write: four terms and four parameters. The Cabibbo idea implies $b = a$ or, in other words, the same Cabibbo angle for the $\Delta S = 0$ and the $\Delta S = 1$ axial currents; of course the expression of the interaction written above is simply the non relativistic part of:

$$\frac{G}{\sqrt{2}} \; J_\mu^{(quark)} \cdot J_\mu^{(lepton)}$$

with

$$J_\mu^{(quark)} = \cos\theta \left( \bar{p}\, \gamma_\mu (1 + a\gamma_5) n \right) + \sin\theta \left( \bar{n}\, \gamma_\mu (1 + b\gamma_5) \lambda \right) \qquad (2)$$

I remember the unconvinced look in the audience, in the early times of the non relativistic quark model when, after writing this current (2) I stated that, in order to give to $g_A/g_V$ for the nucleon the correct value $\approx 1.2$, <u>a</u> had to be $\approx 0.74$[12].

Why should not the quark current be $\bar{p}\gamma_\mu(1 + \gamma_5)n$? And how could the above current, with $a \neq 1$, be reconciled with the current algebra? Only very few people were satisfied when I explained that the interaction written above was an effective interaction or stated differently an element of the S operator; that the quarks concerned were real bound quarks, each surrounded by its $q\bar{q}$ cloud; that, finally, the same question might have been previously asked in connection with the anomalous magnetic moment of the proton. Indeed a massive quark must have, to explain the absolute value of the magnetic moment of the proton, a large anomalous magnetic moment and this produces an effective current $J_\mu^{e.m} = \lambda \bar{\Psi} \sigma_{\mu\nu} q_\nu \Psi$ which certainly does not satisfy the current algebra.

Clearly the fact that with the choice

$$a = b = 0.74 \tag{3}$$

all the semileptonic decays of Cabibbo were reproduced with the $SU_6$ wave functions; and the additional gift or prediction:

$$D/F = \frac{3}{2} \tag{4}$$

was considered gratifying; still I distinctly remember that it was not easy to persuade many people that the current in (2) had no reason to contain $1+\gamma_5$ but that the axial part could well be renormalized, the difference from 1 to 0.7 being ascribable to renormalization effects (the pion ($q\bar{q}$ pairs) around the quark).

Apparently this obvious point of view which has almost always been either expressed or implied in the realistic quark model since its beginning - in full analogy with nuclear physics - was only widely accepted when the words "constituent" and "current" quarks were coined; these words are however perfectly equivalent to "dressed" bound and "bare" quarks which were used previously so that one has in fact an unnecessary duplication in language.

Anyway we must be grateful to the new words if now nobody objects that the weak current, expressed in terms of the constituent quarks, can have the expression (2) with $a \neq 1$; and that the electromagnetic interaction of <u>the constituent quark</u> can have the form:

$$Q \bar{\Psi} \gamma_\mu \Psi A_\mu + \lambda \bar{\Psi} \sigma_{\mu\nu} \Psi F_{\mu\nu} \tag{5}$$

with $\lambda \neq 0$.

## 5. A digression on the mass of the quark

Before going on I would like to make, in this spirit of an attempt to a conceptual clarification, a short remark on the mass of the bound quark. This is rather academic, because, in most of the phenomena, the effective mass of the quark does not really intervene explicitely.

Of course if the free quark exists it has a high value of the mass; otherwise it would have been seen already[13]; as to the mass of the quark inside an hadron, it is, in my opinion at present a matter of convenience to take it large or small; we do not know which is the best description. Certainly the fundamental characteristic of the quark model is that it is a description in terms of a fixed number of constituents. In this description, similar to nuclear physics but, possibly, with very large binding energy, the parameters appropriate to ensure the necessary flexibility can be introduced in several ways; and I do not object if one wishes to introduce these parameters using four component spinors rather than two component spinors. It seems to me however that an important zero order simplification has been to assume that quarks inside an hadron move slowly; to achieve this it was necessary to think to the bound- quark as having large mass.

Why do then so many people prefer to speak in terms of a small quark effective mass (say 1/3 of the nucleon mass)? Of course part of the reason lies, obviously, in the parton picture; but even in the frame of the "low energy" quark model many people prefer to think in terms of small effective mass. Leaving out all arguments based on chiral symmetry I believe that there are, or there have been, two reasons for this. The first has its origin in the confusion, described above, between current and constituent quarks. If the electromagnetic current (5) has to contain only the first term -as it was considered necessary when the constituent and current quarks were identified- then the effective mass of the quark has to be $\sim M_P/3$; indeed this is the only way to obtain a value $\sim 3\,\frac{e}{2M_P}$ for the magnetic moment of the proton (assumed to be in an $L = 0$ state).

The second reason is more subtle; it goes as follows: no matter what the detailed form of the potential well keeping the quarks together is, the relationship between distance $\Delta E$ from level to level and radius $a$ of the potential well is:

$$\Delta E \approx \frac{\hbar^2}{m_q a^2} \qquad \text{or} \qquad a \approx \sqrt{\frac{\hbar^2}{m_q \Delta E}} \tag{6}$$

If, now, $a$ is identified with the electromagnetic radius of our object we have $a \approx (2m_\pi)^{-1}$; taking typically for $\Delta E$ the value $2m_\pi$ we get:

$$\frac{1}{2m_\pi} = \frac{1}{\sqrt{2m_q m_\pi}} \tag{7}$$

so that, again, $m_q \approx 2m_\pi \approx 300$ MeV.

The above deduction assumes, however, that the quark itself is a point quark and that the electromagnetic radius of the hadron has the same order of magnitude as the radius of the potential well keeping the quarks together. If $a$ is much smaller than the e.m. radius of the hadron, the effective mass of the quark can be much larger than $2m_\pi$. In this case, however, the electromagnetic radius of the hadron is practically the same as the electromagnetic radius of the quark itself and the electromagnetic "clouds" of the quarks overlap. This may or may not (depending on the density of these clouds) be a difficulty for electromagnetic additivity.

## 6. A comment on the Melosh-like transformations

After this digression on the mass of the quark we come back to the structure of the currents in terms of the constituent quarks. Consider the matrix element $M_{AB}$ of a local current between two hadronic states; here the current J(X) is a bare current, satisfying the current algebra:

$$M_{AB} = \langle \psi_A(\underset{\sim}{P}) | J(X) | \psi_B(\underset{\sim}{P}') \rangle$$

and $\psi_A$, $\psi_B$ are real hadronic states, moving with momentum $\underset{\sim}{P}$ and respectively $\underset{\sim}{P}'$. In terms of the bare (current) quarks both $\psi_A$ and $\psi_B$ can be written as a superposition of Fock states of the form (for a baryon),

$$\psi = \sum_{\substack{p_1 p_2 p_3 \\ r s t}} f_{rst}(p_1 p_2 p_3)\, a^+_{p_1 r}\, a^+_{p_2 s}\, a^+_{p_3 t} |0\rangle + \sum_{\substack{p_1 p_2 p_3 p_4 p_5 \\ r s t u v}} f'_{rstuv}(p_1 p_2 p_3 p_4 p_5)\, a^+_{p_1 r} \ldots a^+_{p_5 v} |0\rangle + \ldots\ldots \tag{8}$$

In this superposition the first three quark term will be indicated by $\Phi$ and called, for convenience, the model state[14].

To clarify the expression (8) note that, in just the same way, an Hilbert vector for a nucleus composed, say, of three constituent nucleons, has a complicated expression when expressed in terms of bare nucleons and bare pions. Now one can assume that it is possible to define a unitary operator V transforming the (normalized) state $\Phi$ of the bare quarks into the normalized state (8) (which we may call the state of the constituent quarks).

$$\psi_P = V\, \Phi_P \tag{9}$$

Whether this operator V exists and how it can be constructed has been for years (and in a sense still is) the fundamental problem of field theory; we shall add a few comments on it in a moment but, before this, note that using a relation like (9) the matrix element $M_{AB}$ above can be written:

$$M_{AB} = \langle \Phi_A(\underset{\sim}{P}) V^+ | J(X) | V \Phi_B(\underset{\sim}{P}') \rangle =$$

$$= \langle \Phi_A(\underset{\sim}{P}) | V^+ J(X) V | \Phi_B(\underset{\sim}{P}') \rangle \tag{10}$$

In the first form of $M_{AB}$ (10) J(X) is an element of the current algebra (a free quark current) and the states are states of bound dressed quarks (constituent states, as we have already called them); in the equivalent second form the states (model states) are the traditional ones having the usual $SU_6$ properties (at rest) and the current operator $\tilde{J}(X)$:

$$\tilde{J}(X) = V^+ J(X) V \tag{11}$$

is a complicated constituent quark operator.

If J(X) has a structure such as:

$$\bar{\psi}(x)\,\Gamma\,\psi(x) \tag{12}$$

the corresponding expression for $\tilde{J}(X)$ can be tremendously complicated.

Let us now, for simplicity, begin to consider an hadron at rest so that in (10) $\underset{\sim}{P} = \underset{\sim}{P}' = 0$. Of course a factor of the transformation V, let us call if $V_1$, is simply a generator which, for a free quark, separates the positive and negative energies:

$$V_1^+ (\underset{\sim}{\alpha} \cdot \underset{\sim}{p} + \beta m) V_1 = \beta \sqrt{p^2 + m^2} \approx \beta \left( m + \frac{p^2}{2m} \right)$$

The fact that this Tani-Foldy-Wouthuysen transformation or a similar one has to be performed if one likes to give a physical meaning to the position and spin observables, and if one likes to use Pauli spinors instead of Dirac ones, is clear; and, in my opinion, this has always been implied by the users of the non relativistic quark model when interpreting the physical meaning of the variables which they were using. It is my understanding of the situation that the so called Melosh transformation, connecting in the intentions of its Author, the current and the constituent quarks is very much related to this Foldy-Wouthuysen transformation $V_1$; note that the use of an infinite momentum system does not appear to be essential for the criticism which follows.

My objection against taking too seriously this Melosh transformation is that it is only one factor, one tiny factor, of the truth; in other words if we write V in (9) as $V_1 \cdot V_2$, the product of two factors and use only the Melosh operator $V_1$ forgetting about the complicated operator $V_2$ which dresses the quarks and is responsible for the exchange currents and many body effects, we remain in my opinion very far from the correct current $\tilde{J}(X)$. If it is so there seems to be no advantage nor justification in using the Melosh transformation; indeed it seems to me preferable and, I would say necessary, in this situation, to continue to use the usual prescription which we have used since the beginning of the non relativistic quark model; namely to write the simplest expression for the effective current and complicate it by the addition of terms (satisfying the general invariance and conservation conditions) if and when the simplest expression is insufficient.

Coming back to the point which introduced this discussion I consider the factor 0.74 in front of the weak axial current as an example of this approach: namely a renormalization effect of the axial weak hadronic current. In doing so we modify the expression of the weak axial current with respect to the $1 + \gamma_5$ expression for the bare quarks; the modification is both expected and simple: it is expected because, on the basis of the very general treatment contained in the equations (10) and (11), the parameter a in the equation (2) is necessarily different from one; it is therefore natural to attribute to it a value (0.74) which, without additional complications, reproduces correctly the semileptonic decays. It is a simple modification because it is a change in the properties of the individual dressed quark with respect to the individual bare quark and is threfore a one body (that is additive) effect.

I know, of course, that the situation is complicated: we have freedom in choosing the wave functions (pure $SU_6$ representation or mixture between different $SU_6$ representations?); we have freedom in selecting the expression of the current, etc.; my prescription (which, I know, is different from that of other people[15]) amounts to use, at least for the lowest supermultiplet, a pure $SU_6$ representation and modify the current with respect to the bare current form. Of course it would be interesting to find experimentally additional checks of this prescription and ways to distinguish it from other prescriptions.

To conclude these remarks on the relationship between current and constituent quarks, I have stated before that the real problem should be that of determining the complete operator V.

This problem has of course a long story and in particular it was treated at length many years ago in the frame of the Tamm Dancoff and similar methods. The drawbacks of such methods are three: a) they are, essentially weak coupling methods; b) they are non covariant and c) therefore the renormalization procedures are not possible; if these long standing difficulties could be eliminated the determination of V as well as the knowledge, of course, of the approximate form of the unperturbed hamiltonian would allow a complete determination of the various matrix elements also in the general case in which the two states which bracket the current move with different momentum. But, so far, this line has not been followed and the question of writing "wave functions" in motion has been attacked differently.

## 7. Covariant calculations

A long standing and vital problem, in the quark model, or more generally in the dynamics of composite systems, is the one of having a covariant description; much effort has been dedicated to it but in spite of this I feel that an acceptable covariant dynamics of a composite system (with a fixed number of bound objects) has still to be constructed. To try to understand the difficulties I have considered a restricted problem, consisting in studying the relativistic motion of a system which, in its center of mass frame, has a non relativistic internal motion (like e.g. positronium). Even this is not easy as I am going to discuss in my next lecture.[16]

As far as the general problem is concerned there have been in recent years several attempts ranging from ad hoc prescriptions such as those by Feynman et al.[17] to a systematic treatment of the vertices by the Bethe-Salpeter equation. Professor Joos will speak in detail of this here. Another approach will be considered by professor Moorhouse.

There is only one general comment which I want to make here with respect to the covariant treatments; this is that unfortunately the introduction of covariance (which is, I insist, an essential requirement) increases very much the number of free parameters; this is true both if one simply writes down the most general covariant wave function without deducing this wave function from some equation[18]; or if one writes down the B.S. equation and solves it; in one case the free parameters are a necessary feature of the flexibility of the wave function; in the other they are associated to the flexibility of the kernel of the B.S. equation. The reason for this larger number of parameters (with respect to a non relativistic treatment) is easy to understand; it is precisely the same reason why in a non relativistic treatment of the $\beta$ decay the nuclear matrix elements are all reduced (for the allowed transitions) to either a Fermi or a Gamow Teller matrix element; whereas a fully relativistic treatment has 5 or 10 different matrix elements.

## 8. The quark and the parton model

A most important open problem consists in understanding the relationship between the low energy quark model, to which we have confined our attention so far, and the model used to describe the deep inelastic scattering of leptons, in short the parton model. The parton model, no matter how is presented, has two obvious defects: a) that partons are treated as free but cannot be produced; b) that the model is clearly non covariant, all distributions in probability referring to the infinite momentum frame.

In spite of these heavy defects the model has several attractive features; I will leave aside the explanation of scaling which was the original motivation for the model (and probably the parton model is not unique in explaining scaling as the work by Domokos et al.[19] has shown). The particularly interesting feature of the parton model is that through a comparison of the $\ell$ and $\nu, \bar{\nu}$ inclusive form factors one can conclude that partons are fractionally charged with the classical charges 2/3, -1/3. As I repeat this statement refers to <u>partons</u>. Does this imply that we have here an independent demonstration of the attribution of fractionary charges[20] to the quarks which appear in the "low energy" quark model? In other words which is the relation between a moving parton and a static quark? We are, essentially back to problems dealt a moment ago.

Also: is it possible to reconcile the small number of antipartons (with x say $< \frac{1}{10}$) in the parton model with the fact that in the quark model each constituent quark is accompanied by a cloud of $q\bar{q}$ pairs?

Finally, concerning the parton model, I have a specific question which arises out of pure ignorance; I happened to hear at the Erice School in July that there is definite experimental evidence from the inclusive bremsstrahlung by $e^-$ and $e^+$ on protons concerning the fractionary charges of the partons. Does some one have information on this?

I stop here; the survey lectures by professors Cabibbo and Close will deal with these questions; they certainly will contribute in clarifying the connection between the rest quark model and the parton model. The quark model of a baryon being a definite aggregate of three dressed interacting quarks, the parton model should simply be its kinematical transform; the words kinematical transform are however fast to pronounce but difficult to implement[21)].

## 9. The totally open problems

We now come (and I list them purposedly at the end) to the long standing, difficult and entirely open problems; also many of the problems considered so far are open, but the ones I am going to list are perhaps, more open or, stated differently, still more obscure; when they will become clear we shall probably know what exactly the quark is. They are:

1) How many quarks
2) Saturation (and confinement).

How many quarks? Here I am not referring to the question of whether a fourth charmed quark (p' or c) exists; this is a question which will be decided probably soon by the experiments; many experimental groups are at present looking for charmed hadrons in various ways. I am referring instead to the question if the n, p, $\lambda$ quarks (or the n, p, $\lambda$, p' quarks) are "coloured", each appearing in several different varieties. I confine here to list the reasons which are usually stated as suggestive of an internal ("colour") degree of freedom for the quarks, taking at least three values: these reasons are:

1) To have integrally charged quarks behaving in "low energy" phenomena as if they had fractional charge.
2) To have baryons with nodeless space wave functions in the lowest states.

3) To explain the rate of the $\pi_0 \to \gamma\gamma$ decay. According to the Adler-Bell-Jackiw anomaly calculations, agreement is related to the number of quarks.

4) To obtain a reasonable value for the asymptotic behaviour of the ratio $R = (e^+e^- \to \text{hadrons})/(e^+e^- \to \mu^+\mu^-)$; if asymptotically the value of R is related to the sum of the squares of the charges of the quarks by the famous relation:

$$R_\infty = \sum_i Q_i^2 \quad \text{(fermions)} \tag{13}$$

the experimental number at 5 GeV is impossible to achieve without having a conspicuous number of quarks. One triplet of fractionally charged quarks gives $\sum Q_i^2 = 2/3$; three triplets of the Han-Nambu type give 4.

5) To get saturation (and, if one likes, confinement).

Before commenting briefly on the above motivations, I would like to make the following general remark: also if we have only three or four quarks (the 3 Gell Mann - Zweig plus, possibly, p') it is not obvious that they are particles in the usual sense of the word (with asymptotic fields): it is quite possible that quarks (also if only three or four) should be conceived - in a way as yet obscure - as isomorphic to some collective degrees of freedom of hadronic matter (as we did remark long time ago, when commenting on some possible analogies with the Goldhaber-Teller excitations in nuclei[22]). What is difficult to achieve is to find the proper way to assign to such degrees of freedom the quantum numbers typical of quarks (in particular half integer spin). However if the number of quarks is only three - or four - there is also perhaps some chance that they can be particles in the usual sense of the word: particles which the quark hunters may one day find.

But if the number of necessary quarks increases to, say, twelve (through the addition of colour) it looks somewhat unpleasant to conceive all these twelve quark fields as independent asymptotic fields; practically one would then be compelled to think of them in a new way. Therefore to be able to decide, on the basis of the empirical evidence if the introduction of colour is strictly necessary is, in this respect, extremely important. The question is therefore; how compelling are the arguments 1) to 5) in the list given above. I will briefly express my point of view on each of them noting, however, that for the point 2) and especially 4) I do not share the confidence held by the majority of the theorists.

1) The requirement in question has been the motivation of the Han-Nambu three triplet scheme[20). It is not a necessary requirement - at the present state of our knowledge - but can be aesthetically appealing.

2) It is often ignored that Fermi statistics (for baryons) can be satisfied also with one triplet of quarks only; an $L=0$ spatially antisymmetric wave function can be easily constructed. It is possible to depress in energy the states with space antisymmetric wave function by the assumption of Majorana exchange forces between quarks, attractive in antisymmetric states, repulsive in symmetric; but it is an open dynamical problem[23) to understand if this is compatible or not with the observed baryon spectrum. Note finally that an antisymmetric space wave function is not necessarily incompatible with the electromagnetic form factor of the nucleon.

3) How strong is the argument from the $\pi^0 \rightarrow \gamma\gamma$ decay a demonstration of the necessity of three triplets I do not know; I

only refer to a paper by Drell for an unconventional discussion of this question[24).]

4) $R_{\infty}$: The equation (13) is only true in some-asymptotically free-theories; moreover it looks to me very naive to believe (each year) that we are asymptotic. In other words: I may envy but do not share the sort of faith which so many people have in the equation (13).

I should now proceed to the question of saturation and - possibly - of confinement both in the various colour models and in the different schemes (bag models etc.). However the variety of ideas which are in circulation here would make any sensible presentation too long and not in line with the character of an introductory talk I tried to keep. Therefore I prefer to stop here also because professor Nambu and dr. Giles may consider, from different points of view, these questions in their lectures.

In concluding I feel that I must apologize for two different reasons: in the first place for having sometimes been perhaps too sharp in my assertions in what is undoubtely a controversial subject; in the second place for having dedicated practically this whole introduction to conceptual and sometimes academic questions; of course one single unambiguous prediction of a number has much more value than thousands of words.

## Footnotes and References

1) If the interpretation in terms of charmed quarks is the correct one, perhaps the quark hunters may receive some hope from the fact that the effective mass of these charmed

quarks, when bound, is as high as 2 GeV. The free charmed quark must then have a mass larger than this; to think in terms of heavy normal quarks (with mass, say, 5 or 10 GeV) strongly bound as suggested when the realistic non relativistic quark model was first introduced (G. Morpurgo, Physics N.Y. 2, 95 (1965)) does not seem so impossible once this new heavy scale has been set. Of course this remark implies that quarks are particles and not something else (compare the comments in Sect. 9).

2) A.P. French, Mesons -1975 (CERN preprint 75-38), Invited talk at the Palermo June 1975 Conference of the EPS.

3) G. Morpurgo: a) Lectures on the quark model (Erice lectures 1968) in "Theory and phenomenology in particle physics" part A, edited by A. Zichichi, p. 84-217 (Academic Press 1969); b) Rapporteur talk on the Quark Model at the XIV International Conference on high energy physics, Vienna 1968 (Proceedings, p. 225-252).

4) The importance of the relativistic and exchange corrections to the magnetic moment of the deuteron and of the light nuclei is still an open question often discussed in the literature.

5) C. Becchi and G. Morpurgo, Phys. Rev. 140B, 179 (1965).

6) A new measurement has recently been made by the Orsay group (G. Cosme et al. L.A.L. preprint 1279, August 1975) giving for the branching ratio of the $\phi$ into $\eta\gamma$ a value $(1.5 \pm 0.4)\%$ to be compared with a previous value by the same group of $(2.6 \pm 0.7)\%$ and with a value of $(7.3 \pm 0.8)\%$ by the CERN-Bologna collaboration.

7) ($\bar{K}^*_o \to \bar{K}_o + \gamma$): W.C. Carithers et al., University of Rochester preprint UR-525, April 1975; ( $\rho^- \to \pi^- + \gamma$ ): B. Gobbi et al., Phys. Rev. Lett. 33, 1450 (1974).

8) R.H. Dalitz and D. Sutherland, Phys. Rev. 146, 1180 (1966).

9) C. Becchi and G. Morpurgo, Phys. Letters 17, 352 (1965).

10) R.G. Moorhouse, Phys. Rev. Lett. 16, 772 (1966).

11) L. Copley, G. Karl and E. Obryk, Nucl. Phys. B 13, 303 (1969); id. Phys. Letters, 29B, 117 (1969).

12) Compare in ref. (3a) above the presentation at p. 206 and the discussion at page 203.

13) This is the old fashioned attitude; different confinement mechanisms are in fact being actively discussed.

14) Compare the sections 8.1, 8.2 and 8.3 in ref. (3a) above; compare also J.S. Bell and H. Ruegg, these Proceedings.

15) For instance of the Orsay group (compare A. Le Yaouanc et al. LPTHE 75/11 and the lecture to be given here by dr. Oliver); a still different and widely adopted attitude is that exemplified by dr. A. Hey - these Proceedings.

16) G. Morpurgo, Separation of kinematics and dynamics in the relativistic treatment of composite systems (Proceedings of the Erice School of Subnuclear Physics, 1974).

17) R.P. Feynman et al., Phys. Rev. D3, 2706 (1971).

18) G. Morpurgo, The present status of the radiative decays of mesons, in "Properties of the fundamental interactions" ed. by A. Zichichi (published by Compositori-Bologna)-Proceed. of the ninth course (1971) of the International School of Subnuclear Physics, p. 432-477.

19) G. Domokos et al. Phys. Rev. D3, 1184 and 1191 (1971).

20) To be precise one should say: of effective fractionary charges; indeed if a model of the Han Nambu type (Phys.Rev. 139B, 1006 (1965) and Phys. Rev. D10, 674 (1974)) holds, with integer charges, an effective fractional charge behaviour continues to be seen up to the threshold of colour for real processes; and for virtual processes the modifications sufficiently below this threshold can be comparatively small.

21) This is due to the fact that relativistically is far from easy to separate the kinematics from the internal dynamics.

22) Compare ref. (3a) p. 210.

23) Compare ref. (3a) Sect. 6.4.

24) S.D. Drell, Phys. Rev. D7, 2190 (1973).

# ELECTROMAGNETIC TRANSITIONS OF NUCLEON RESONANCES, CONSTITUENT QUARK THEORY, AND THE MELOSH-GILMAN-KUGLER-MESHKOV PARAMETRIZATION

R.G. Moorhouse

Department of Natural Philosophy, Glasgow University, Scotland

I have been asked to give this background talk to the Workshop since the accumulation of data and partial wave analysis of radiative transitions of nucleon resonances gives one of our most important insights into the constituent quark model. Also, one of the topics of the Workshop is the relationship between the constituent quark model and the Melosh transformation and, because of the abundance of data, there has been much phenomenological fitting of the parameters that occur in the simplified version of the Melosh transformation proposed by Gilman, Kugler and Meshkov[1] (this version being nearly identical with "$\ell$-broken $SU(6)_W$" theory[2]). Consequently radiative nucleon transitions form an area where one can compare constituent quark model theory and the Melosh-Gilman-Kugler-Meshkov parametrization, leading hopefully to the improvement of both. This talk also deals with radiative transitions with virtual photons, $\gamma_v + N \rightarrow N^*$ (coming from data on $e^- + N \rightarrow e^- + N + \pi$) where the Melosh transformation is silent but the existing quark model has something to say, partly right and partly wrong.

## $N^* \rightarrow N\gamma$ AMPLITUDES FROM PARTIAL WAVE ANALYSIS OF $\gamma N \rightarrow \pi N$

The principal object of these analyses is to obtain the $(N^*, N\gamma)$ coupling constants for the various $N^*$ resonances. There have been many results in this field in the past two to three years, largely based on the development of the technique of fixed-t dispersion relations. The method[3] is to parametrize the imaginary part only of the amplitudes using a partial wave expression for $\gamma N \to \pi N$, to find the real part from these imaginary parts by fixed-t dispersion relations, and then find the parameters of the imaginary parts by fitting the whole amplitude -- imaginary part and the resulting real part -- to the data. The method is powerful because i) the imaginary part of the amplitude is resonance dominated in the resonance region; ii) the masses and widths of the majority of the most important resonances in the resonance region are already known and fixed from $\pi N \to \pi N$ partial wave analysis and so largely pre-determine the form of the imaginary part with the $N^* N\gamma$ coupling constants always as the free parameters; iii) it connects the amplitudes for $\gamma p \to \pi^+ n$ and $\gamma n \to \pi^- p$ through the fixed-t dispersion relations. A formal weakness is that the summation of the partial wave expansion is not certainly known to be convergent over the t-region used, though it can be reasonably conjectured to be good on the basis of the Mandelstam representation from $-t = 0$ to $-t = 1$ $(\mathrm{GeV/c})^2$, and the method has been used with apparent success as far as $-t = 1.6$ $(\mathrm{GeV/c})^2$; this point deserves and will probably receive further study through techniques of analytic continuation.

Recently Devenish, Lyth and Rankin[4)] have extended the method by also parametrizing the imaginary part of the high-energy amplitudes in a Regge-like manner, and simultaneously fitting both the higher energy and the resonance region data; there is of course a contribution to the resonance region real part from the high-energy imaginary part and thus the high-energy data (presumably) help to fix the resonance parameters. In a submission to the Palermo conference Barbour and Crawford[5)] have similarly fitted both the resonance region and the high-energy data, obtaining a better fit than that of Devenish et al. (though Barbour and Crawford have at the moment a gap in the data fitted in the energy region $2.0\ \mathrm{GeV/c^2} < E_{cm} < 2.6\ \mathrm{GeV/c^2}$). It suffices for the moment to say that this most recent analysis contains no surprises or upsets concerning those resonance couplings which were previously reasonably well determined; it does depart from previous analyses in having larger couplings of the $N^{*+}$ resonance, $p^*_{13}(1800)$ and the magnitudes and signs of the $p_{13}$ amplitudes happen to agree with naive quark model calculations and have some general significance, as we shall see.

For a given charge state of $N^*$, there are in general two independent amplitudes for the decay $N^* \to \gamma N$ of a resonance of spin J. In the $N^*$ centre of mass the $\gamma$ and N come off in opposite directions along say the z-axis and the usual amplitudes are helicity amplitudes in which the z-component of spin is 1/2 or 3/2. (These of course are linear superpositions of the electric and magnetic multipole amplitudes.)

$$F_{1/2}(N^* \to \gamma N): \quad \underset{\gamma}{\longleftarrow} \overset{S_z^{N^*} = 1/2}{N^*} \underset{N}{\longrightarrow} \qquad S_z^{\gamma} = 1 \quad S_z^{N} = -1/2$$

$$F_{3/2}(N^* \to \gamma N): \quad \underset{\gamma}{\longleftarrow} \overset{S_z^{N^*} = 3/2}{N^*} \underset{N}{\longrightarrow} \qquad S_z^{\gamma} = 1 \quad S_z^{N} = 1/2$$

Since the partial wave analysis actually determines a product $F(\pi N \leftarrow N^* \leftarrow \gamma N) = F(\pi N \leftarrow N^*)\ F(N^* \to \gamma N)$ of which we usually know the magnitude of $F(\pi N \leftarrow N^*)$ from $\pi N$ elastic partial wave analysis, we usually define and tabulate

$$A_{1/2}(N^* \to \gamma N) = F_{1/2}(N^* \to \gamma N) . \text{ sign } (N^* \to \pi N)$$

$$A_{3/2}(N^* \to \gamma N) = F_{3/2}(N^* \to \gamma N) . \text{ sign } (N^* \to \pi N) .$$

Also for <u>isospin-1/2</u> resonances the amplitudes for proton states and neutron states are independent; so we have four amplitudes for each isospin-1/2 resonance $A^p_{1/2}$, $A^p_{3/2}$, $A^n_{1/2}$, $A^n_{3/2}$ -- and of course two $A_{1/2}$, $A_{3/2}$ for each isospin-3/2 resonance. It is well known that the <u>signs</u> and magnitudes of these amplitudes, being not determined by SU(3), contain much information crucial to models. A few of the important features are:

a) dominance of the magnetic multipole amplitude[6] for the resonances of the $\Delta$-Regge recurrence

| | $\{56\}L = 0^+$ $p_{33}(1236)$ | $\{56\}L = 2^+$ $f_{37}(1950)$ | $\{56\}L = 4^+$ $h_{3,11}(2400)$ |
|---|---|---|---|
| $\dfrac{\text{electric multipole}}{\text{magnetic multipole}}$ : | $< 0.02$ | $< 0.07$ | $< 0.18$ ? |

naive constituent quark models give electric multipole = 0 for all these resonances.

b) the prominent resonances of the $\{70\}L = 1^-$ multiplet:

i) $d_{13}(1520)$ $A^p_{3/2}$ large +, $A^n_{3/2}$ large −, $A^p_{1/2}$ small

ii) $d_{15}(1670)$ $A^p_{1/2}$, $A^p_{3/2}$ small -

iii) $d_{33}(1670)$ $A_{1/2}$, $A_{3/2}$ medium -, large +

These qualitative features agree with naive constituent quark models.

## SINGLE QUARK TRANSITION MULTIPOLES

Many of the qualitative features can be succintly expressed using the electric and magnetic multipoles appropriate to the single quark transitions from $N^*$ to N

$N^*$, orbital angular momentum L

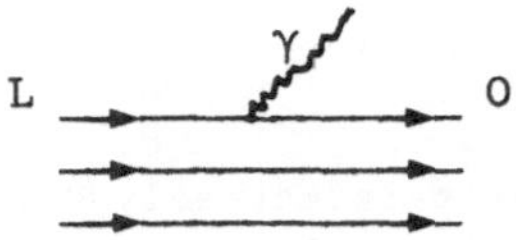


N, orbital angular momentum 0 .

Naive constituent quark calculations de-excite the $N^*$, of total quark orbital angular momentum, L, by photon emission involving a single quark; thus the orbital angular momentum L resides on that quark, and the photon is emitted in an M(L - 1), M(L + 1), or EL pole state. The electronic multipole emission can take place either through the quark charge when we denote it by EL′ or through the quark magnetic moment when we denote it by EL″. In a recent work, using three different partial wave analyses[4,7,8] of pion photoproduction Babcock and Rosner[9] verify that the following single quark transition multipoles are dominant:

$$\{70\}L = 1^- \text{ multiplet: } E1', M2$$
$$\{70\}L = 2^+ \text{ multiplet: } E2', M3 .$$

This qualitative observation is in agreement with naive quark model calculations, as we should expect from the previously noted agreements.

## MELOSH TRANSFORMATION PARAMETRIZATION AND NAIVE QUARK MODEL CALCULATIONS

In fact Babcock and Rosner carry out their investigation in the context of the (Melosh transformation + single quark transition) hypothesis developed by many authors[1,2,10] and their single quark multipole analysis follows earlier work of Gilman and Karliner[11]. We remember that the Melosh transformation[12] work following ideas of Gell-Mann distinguish between current quarks and constituent quarks, states that hadrons are in irreductible representations of $SU(6)_W$ of constituents associated with orbital angular momentum (that is $\{70\}L = 1^-$, $\{56\}L = 2^+$, etc., as in the naive constituent quark model)

$$|\text{hadron}\rangle = |\text{Irred. Rep., constituent quarks}\rangle$$

that the Melosh transformation is hypothesised to connect current quark representations to constituent quark representations

$$|\text{hadron}\rangle = V|\text{Irred. Rep., current quarks}\rangle$$

that matrix elements of interest have a current, simply expressible in irreducible representations of current quarks, sandwiched between hadrons so that

axial vector current ($\pi$'s) → $Q_5$; photon current → $D_+$

$$\langle \text{hadron}^* | (Q_5 \text{ or } D_+) | \text{hadron} \rangle =$$

$$= \langle \text{Irred. Rep. currents}^* | V^+ \; (Q_5 \text{ or } D_+) V | \text{Irred. Rep. currents} \rangle$$

that $\tilde{Q}_5 = V^+ Q_5 V$ and $\tilde{D}_+ = V^+ D_+ V$ have relatively simple properties under the "current quarks" group and that they satisfy the single quark transition hypothesis. This leads for all $N^*$'s in a given supermultiplet such as $\{56\}L = 2^+$, to the expressibility of transition amplitudes

| | | |
|---|---|---|
| $F(N^* \rightleftarrows \pi N)$ | in terms of two independent amplitudes (that is per supermultiplet) | say P (P-wave) and F (F-wave) |
| $F(N^* \rightleftarrows \gamma N)$ | in terms of four independent amplitudes (per supermultiplet) | A, B, C, D |

These different amplitudes, P, F, A, B, C, D, can of course be identified with terms which do, or could (see below), occur in naive quark model calculations; only in such naive quark model calculations the terms vary somewhat within the supermultiplet and are predicted. The operational content of the Melosh-Gilman-Kugler-Meshkov parametrization is the simple single quark transition hypothesis, making it a phenomenology, with the restrictive assumption of constancy, within a supermultiplet, of the arbitrary parameters such as P, F, A, ... .

Signs of $N^* \to \pi N$ amplitudes

Now let us concentrate for a moment on the $N^* \to \pi N$ transition amplitudes: as emphasized by Babcock and Rosner[9], and by previous workers in the field, a total knowledge from experiment and PWA of amplitudes

$$F(\gamma N \to N^* \to \pi N) = F(\gamma N \to N^*)\, F(N^* \to \pi N)$$

for all $N^*$ in the $\{56\}L = 2^+$ would lead to a knowledge not only of the relative signs and magnitudes of A, B, C. D, but also of the pionic transition amplitudes P and F. The same applies to the $\{70\}L = 1^-$ supermultiplet, where the corresponding pionic transition amplitudes are S-wave and D-wave. There is independent evidence for both supermultiplets from the $\pi N \to N^* \to \pi\Delta$ transitions. The pionic situation may be summarized as follows:

| | $N^*$ $\{70\}L = 1^-$ | $N^*$ $\{56\}L = 2^+$ |
|---|---|---|
| $\gamma N \to N^* \to \pi N$ | sign(S) = -sign(D) | sign(P) = -sign(F) (from $p_{13}$ only) |
| $\pi N \to N^* \to \pi\Delta$ | sign(S) = -sign(D) | sign(P) = +sign(F) (from $f_{15}$ only) |
| naive constituent quark model | sign(S) = -sign(D) | sign(P) = -sign(F) . |

As a comment on these results, we remark that both the $p_{13}(\sim 1800)$ and the $f_{15}(\sim 1690)$ belong to the $[8,2]L = 2^+$ submultiplet of the $\{56\}L = 2^+$. The $p_{13}$ result comes from a number of analyses of photoproduction, including the very recent one of Barbour and Crawford (as noted above) and the results from those analyses covering this energy region may be listed as shown, in appropriate units

| | Naive Quark model (no free parameters) | Barbour and Crawford[5] | Devenish, Lyth, Rankin[4] | Crawford[13] | Metcalf and Walker[7] | KMORR[8] |
|---|---|---|---|---|---|---|
| $A^p_{1/2}[p_{13}(\sim 1800)]$ | 100 | 105 | 25 ± 34 | 22 ± 12 | 0 ± 25 | 26 |
| $A^p[p_{13}(\sim 1800)]$ | -30 | -60 | -87 ± 57 | -16 ± 16 | 0 ± 22 | -12 |

We see that we must still regard the sign of this photoproduction amplitude as being only dubiously determined, but all present indications are in favour of the naive quark model sign. The $f_{15}$ determination, which is against the quark model sign, comes from the $N^* \to \pi\Delta$ component of $\pi N \to N^* \to \pi\Delta$ in the Berkeley/SLAC analysis of Cashmore et al.[14]. [The

$P_{33}(2000)$ sign has been quoted, from photoproduction, as evidence against the quark model sign; however, this evidence is much more dubious than that in the $P_{13}$ case.] Thus the pionic situation for the {70}L = $1^-$ has good agreement -- and from a number of resonances -- of $\gamma N \to \pi N$ with $\pi N \to \pi\Delta$. However, for the {56}L = $2^+$, "Melosh" is in disagreement, but based in each case on one resonance only. Existing naive quark model calculations would be supported by the sign indicated by photoproduction in the {56}L = $1^-$ experimental results.

$\gamma N \to N^*$ amplitudes

For $N^* \to \gamma N$ there are in general four Melosh transformation parameters: A, B, C, D, for $N^*$ in a given supermultiplet such as the {56}L = $2^+$. Because of angular momentum limitations there are only three parameters, A′, B′, C′ for $N^*$ {70}L = $1^-$. Now existing naive quark model calculations give only A and B type parameters; C and D type parameters would be given by spin-orbit and more complicated coupling terms in the electromagnetic coupling.

Now it had previously been noted by Gilman and Karliner[11)], and in work by Cashmore, Hey and Litchfield[15)] that a best fit to the photoproduction of {70}L = $1^-$ resonances demanded a C′ term. Now in a further very recent paper, Cashmore, Hey and Litchfield[15)] find that both C and D terms are needed in their best fit to the pion photoproduction amplitudes $\gamma N \to N^* \to \pi N$ for $N^*$ {56}L = $2^+$. [The opposite conclusion was reached in earlier work by Gilman and Karliner[11)]]. The indications from this Melosh-type parametrization of spin-orbit coupling terms being needed in electromagnetic transitions in the naive quark model should be taken moderately seriously. However one should warn, as indeed Cashmore, Hey and Litchfield do themselves warn, against taking these numbers too seriously; they do a fit to one partial wave analysis only, which is itself not a perfect fit to the existing experiments. There is obviously a danger of fitting the noise rather than the signal as is shown by the fact that one of the few relatively well determined $N^* \to \gamma N$ amplitudes of L {56}L = $2^+$, the $A_{3/2}$ of $f_{37}(1950)$, badly fails to be fitted by their four free parameter parametrization; indeed the zero free parameter naive quark model does better.

FORM FACTORS

The approximately dipole behaviour of the nucleon form factors $G_E^p(q^2)$, $G_M^p(q^2)$, $G_M^n(q^2)$ is well established. We are also concerned with form factors of a different kind $N \overset{q^2}{\rule{}{}} N^*$, namely the behaviour of the $N^*N\gamma$ vertex as a function of $q^2$, these of course are just the amplitudes (for $q^2 = 0$) whose significance for the quark model and for the Melosh transformation work we have previously discussed. We will now discuss their variation with $q^2$ and we have, through electroproduction experiments of the last 2-3 years, and a fixed-t dispersion relation partial wave analysis submitted to the Palermo Conference by Devenish and Lyth[16] qualitatively better information than before available.

The form factors determined are for positive particles $p \overset{q^2}{\rule{}{}} N^{*+}$ (more explicitly for $ep \to ep\pi^0$) and some of the main qualitative features are -- in the region of experiment $|q^2| < 1.2$, and in terms of electric and magnetic multipoles (using the CGLN notation)

| | |
|---|---|
| $N^{*+} = p_{33}(1236)$ | M1+ not of dipole form |
| | $\lvert E1+/M1+\rvert \simeq 0.02$ for $\lvert q^2\rvert \leq 1.2$ (GeV/c) |
| | the scalar amplitude (longitudinal photons), small, |
| | but finite: $S_{1+}/M_{1+} \sim -0.06$ |
| $N^{*+} = d_{13}(1520)$ | E2- of dipole form |
| | M2- only slowly decreasing |
| | S2- quickly decreases in magnitude, |
| $N^{*+} = S_{11}(1505)$ | E0+ decreases more slowly than dipole |
| | S0+ might be dipole-like, or like E0+ |

(Both the $d_{13}$ and $s_{11}$ belong to the $[8,2]L = 1^-$ submultiplet of the $\{70\}L = 1^-$)

| | |
|---|---|
| $f_{15}(1690)$ | E3- approximately dipole-like |
| | M3- decreases slowly |
| $f_{37}(1950)$ | M3+ probably decreases more slowly than dipole |
| | E3+ small $\lvert E3+/M3+\rvert < 0.04$ . |

Some of these features are illustrated in Fig. 1. We should first of all iterate that even the nucleon form factors are a well-known unsolved problem at any rate for small $|q^2| < 1\text{-}2\ (\mathrm{GeV}/c)^2$. For large $|q^2|$, where asymptotic approximations may be made, there are indications that the quark model may give the $1/q^4$ presumed asymptotic behaviour of nucleon form factors and the $1/q^2$ presumed asymptotic behaviour of pion form factors.

Such approaches do not apply to the range of $|q^2| < 1\text{-}2$ which is now in question, and with a theory, such as the naive constituent quark model, which gives first order correct answers to radiative transitions with $q^2 = 0$, a reasonable expectation would be to develop it to nearby values of $q^2 \neq 0$. Apart from problems in the containment or binding force and the exact form of the wave function in the hadron rest frame, the main technical problem is in boosting the wave function of one of the hadrons to non-zero hadron momentum. The work of Feynman, Kislinger and Ravndal[17)], Ravndal[18)], Lipes[19)], Fujimura et al.[20)], Gonzales and Watson[21)] and of Kellett[22)] has drawn attention to the importance of the Lorentz contraction in developing the dipole form factor. However, besides difficulties of principle with time-like states in those of the models which use four-dimensional oscillators, all seem to founder on technical difficulties associated with the spinor part of the quark wave function, which develops unwanted extra factors in the form factor. Henriques et al.[23)] have shown that it may be possible to avoid these technical difficulties by using more general relativistic spinor wave functions, but only at the expense of giving up the evident generalization from the non-relativistic quark model and SU(6).

One is left with a comparison between form factors and there is now a sharp comment that can be made on the question raised by Close and Gilman[24)] some years ago. The question can easily be seen in the context of the non-relativistic harmonic oscillator model, and it concerns the $d_{13}$ and $f_{15}$ resonances, and is exhibited in its sharpest form by considering the helicity amplitudes $A^p_{1/2}$ and $A^p_{3/2}$, for $N^* \rightarrow \gamma N$, previously defined, for the proton state of these resonances. For the $d_{13}(1520)$, the non-relativistic oscillator model gives for helicities 1/2 and 3/2, respectively

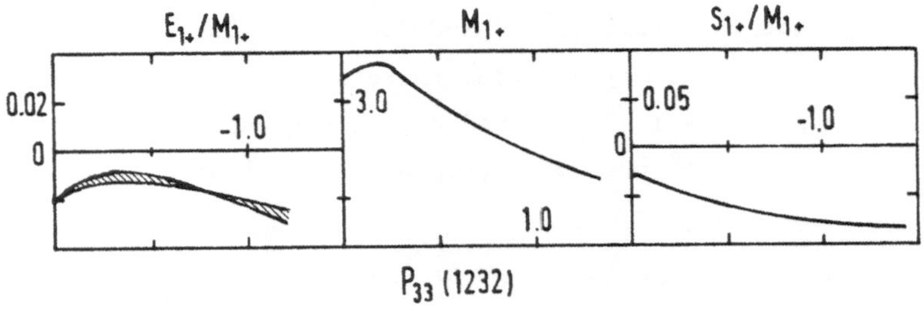


a)

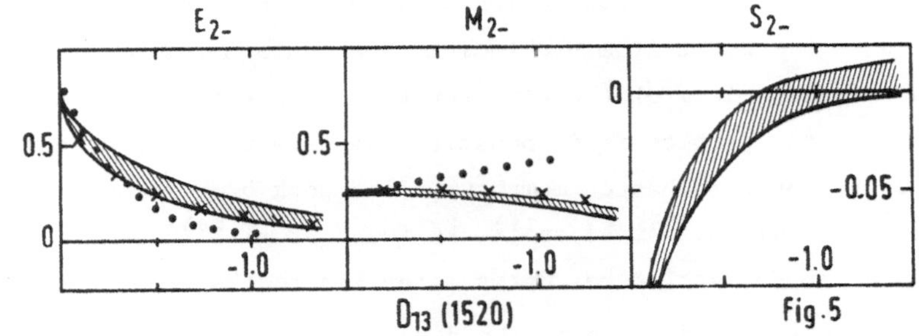


b)

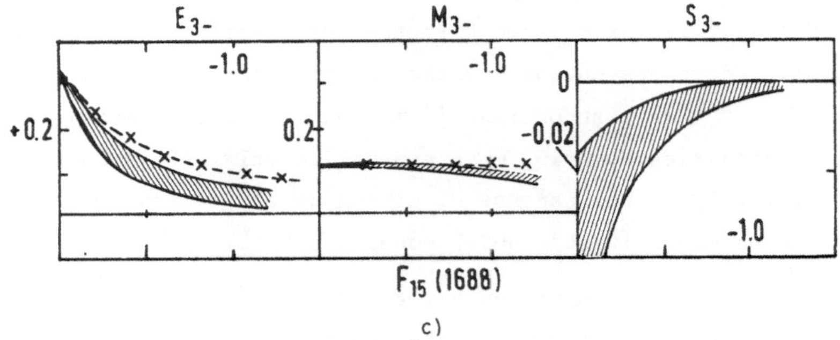


c)

Fig. 1 : The $(PN^{*+}\gamma_V)$ electric multipole $(E_{\ell\pm}, M_{\ell\pm}, S_{\ell\pm})$, magnetic multipole $(M_{\ell\pm})$, and scalar multipole $(S_{\ell\pm})$ form factors for some prominent resonances from partial wave analysis[16] of $\pi^0$ electroproduction, plotted as a function of $q^2$, the four-momentum squared of the virtual photon, from $q^2 = 0$ to $q^2 \simeq -1.3$. For the electric and magnetic multipoles of the $d_{13}$ we show the result, in the dotted lines, of the non-relativistic quark model form factor calculation (normalized, with a common normalization, to the experimental result at $q^2 = 0$).

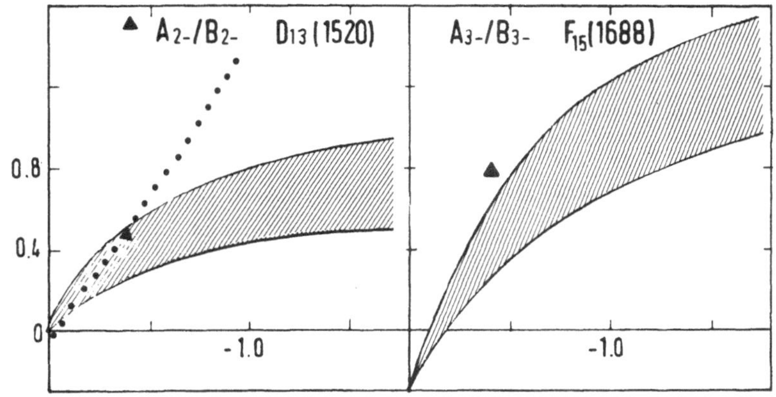


Fig. 2 : The ratio of helicity amplitudes plotted as a function of $q^2$, the four-momentum squared of the virtual photon, for the $d_{13}$ and $f_{15}$ resonances, from partial wave analysis[16] of $\pi^0$ electroproduction. The ratios plotted use the Hebb-Walker amplitudes, and are related to the helicity amplitudes of the text by

$$\frac{A_{2-}}{B_{3-}} = \sqrt{3}\,\frac{A^p_{1/2}}{A^p_{3/2}}\,(d_{13})\ ; \qquad \frac{A_{3-}}{B_{3-}} = 2\,\sqrt{3}\,\frac{A^p_{1/2}}{A^p_{3/2}}\,(f_{15})\ .$$

The figure illustrates the steep rise in the ratio found in the partial wave analysis; the dotted line illustrates, for the $d_{13}$, the corresponding rise predicted by the non-relativistic quark model. This steep rise is related to the very different behaviour of the electric and magnetic multipoles (which are superpositions of the helicity amplitudes) shown in Fig. 1.

$$A^p_{1/2} = a\left[\frac{\underline{k}^2}{(\alpha^2/g)} - 1\right] \exp\left[-\underline{k}^2/6\alpha^2\right]$$

$$A^p_{3/2} = -\sqrt{3}\ a \exp\left[-k^2/6\alpha^2\right] ,$$

where $\underline{k}$ is the photon three-momentum and we use the proton Breit frame, $\alpha$ is the usual harmonic oscillator spring coupling constant, g is the quark gyromagnetic ratio and the constant a contains the quark magnetic moment. As first shown by Copley, Karl and Obryk[25], a value of $\alpha^2/g$ reasonably consistent with that required from other aspects of the quark model leads to a vanishing or small value of $A^p_{1/2}(d_{13})$, and also simultaneously of $A^p_{1/2}(f_{15})$, for real photons, as required by the experiments. Now if we go over from real photons to the virtual photons of electroproduction $\underline{k}^2$ gets larger so $A^p_{1/2}$ increases relative to $A^p_{3/2}$ and at $\underline{k}^2 \simeq 1\ (\text{GeV/c})^2$ for harmonic oscillator wave functions, the $d_{13}$ should be predominantly electroproduced into the $A^p_{3/2}$ state. Of course the exact rate at which this switch occurs depends on the details of the wave functions, and for Coulomb-like wave functions which might be more appropriate for asymptotic freedom theories like that of De Rujula, Glashow and Georgi, the change would take place more slowly.

We see (Fig. 2) that for both the $d_{13}$ and $f_{15}$ amplitudes the experiments and partial wave analysis show that a change of the required character does indeed take place, though at a somewhat slower rate than predicted by the harmonic oscillator wave functions. In this comparison the term $e^{-k^2/6\alpha^2}$, which is the common part of the form factor and may well be inaccurately predicted by the simple model, drops out in the quotient. We have chosen to use helicity amplitudes for the comparison, and these of course can be expressed as a linear superposition of the multipole amplitudes previously plotted. The experimental switch from helicity 1/2 towards helicity 3/2 is expressed in multipole amplitude tersm as a slower decrease of the magnetic amplitude. We now see that this slower decrease is indeed a feature of the theory (Fig. 1).

Melosh transformation work, in its present stage of development, has nothing at all to say about form factors or energy dependence within or between multiplets, but at least one such phenomenological analysis of electroproduction has been made[26] essentially obtaining the parameters A, B, C, D (under various simplifying assumptions) as functions of $q^2$.

## REFERENCES


1) F.J. Gilman, M. Kugler and S. Meshkov, Phys. Letters 45B, 481 (1973); Phys. Rev. D 9, 715 (1974).

2) A.J.C. Hey, J.L. Rosner and J. Weyers, Nuclear Phys. B61, 205 (1973).

3) R.G. Moorhouse and H. Oberlack, *in* Proc. of the XVI Int. Conf. on High-Energy Physics, Chicago-Batavia, Ill., 1972), Vol. I, p. 182.
R.C.E. Devenish, D.H. Lyth and W.A. Rankin, Phys. Letters 47B, 53 (1973).
R.G. Moorhouse, H. Oberlack and A.H. Rosenfeld, Phys. Rev. D 9, 1 (1974).

4) R.C.E. Devenish, D.H. Lyth and W.A. Rankin, Phys. Letters 52B, 227 (1974).

5) I.M. Barbour and R. Crawford, Glasgow preprint (1975).

6) G. Knies, R.G. Moorhouse and H. Oberlack, Phys. Rev. D 9, 2680 (1974).

7) W.J. Metcalf and R.L. Walker, Nuclear Phys. B76, 253 (1974).

8) G. Knies, R.G. Moorhouse, H. Oberlack, A. Rittenberg and A.H. Rosenfeld, Proc. of the XVII Int. Conf. on High-Energy Physics, London 1974, edited by J. Smith (Science Research Council, Chilton), reported by D.H. Lyth, p. 150.

9) J. Babcock and J. Rosner, Multipole Analysis of resonance photoproduction, Cal. Inst. of Technology preprint, CALT-68-485 (1975).

10) F. Gilman, M. Kugler and S. Meshkov, Phys. Letters 45B, 481 (1973); Phys. Rev. D 9, 715 (1974); A.J.G. Hey, J. Rosner and J. Weyers, Nuclear Phys. B61, 205 (1973); F. Gilman and I. Karliner, Phys. Letters 46B, 426 (1973).

11) F. Gilman and I. Karliner, Phys. Rev. D 10, 2194 (1974).

12) H.J. Melosh, IV Thesis, Cal. Inst. of Technology (1973), unpublished; Phys. Rev. D 9, 1095 (1974).

13) R. Crawford, Nuclear Phys. (to be published).

14) R.J. Cashmore et al., SLAC-PUB-1387, LBL 2634 (1974). See also R. Longacre, *in* Proc. Palermo Internat. Conf. on High-Energy Physics, 1975 (to be published).

15) R.J. Cashmore, A.J.G. Hey and P. Litchfield, Further applications of $SU(6)_W$ decay schemes for baryons, preprints (1975).

16) R.C.E. Devenish and D.H. Lyth, Electromagnetic form factors of $N^*$ resonances and their determination from pion electroproduction, DESY report, DESY 75/04.

17) R.P. Feynman et al., Phys. Rev. D 3, 2706 (1971).

18) F. Ravndal, Phys. Rev. D 4, 1466 (1971).

19) R.G. Lipes, Phys. Rev. D 8, 2849 (1972).

20) K. Fujimura et al., Progr. Theoret. Phys. (Kyoto) 43, 73 (1970).

21) M.A. Gonz les and P.J.S. Watson, Nuovo Cimento 12A, 889 (1972).

22) B.H. Kellett, Ann. of Phys. 87, 60 (1974).

23) A.B. Henriques, B.H. Kellett and R.G. Moorhouse, Ann. Phys. (to be published). See also recent work by Meyer (Bonn preprint, 1975).

24) F.E. Close and F.J. Gilman, Phys. Letters 38B, 541 (1972).

25) L.A. Copley, G. Karl and E. Obryk, Nuclear Phys. B13, 303 (1969).

26) C. Avilez and G. Cocho, Phys. Rev. D 10, 3638 (1974).

THE SPECTRUM OF BARYONIC STATES

R. H. Dalitz

Department of Theoretical Physics
Oxford University

## 1. Patterns for the Baryonic States

Our standpoint in these lectures is the three-quark model for the baryonic states. With this model, such states may be classified according to the permutation symmetry of their space wavefunction, as being symmetric (S), antisymmetric (A), or of mixed symmetry (M). We shall assume here that SU(6) symmetry holds for the quark-quark interaction, so that the baryonic states may also be classified by the SU(6) representation to which they belong. The SU(6) representations to which the three-quark system can belong are $\underline{56}$, $\underline{70}$, and $\underline{20}$ dimensional; these may also be characterized by their permutation symmetry, this being S, M and A, in turn.

The empirical fact is that the lowest baryon supermultiplet is a $\underline{56}$ representation with LP=0+. With considerable generality, the symmetry of the ground state space wavefunction is expected to be S, so that the overall wavefunction, including the space, spin and unitary-spin factors, therefore has S permutation symmetry. This is contrary to the Spin-Statistics Theorem of Pauli,[1] and we must conclude that there are further quantum numbers associated with the quark. Following Gell-Mann's proposal,[2] this additional variable is now termed colour. In order to achieve its purpose, that is to provide the required antisymmetric factor in the wavefunction for three quarks, this variable must have three and only three eigenstates for the quark, and these are generally referred to as red, blue and green, the required factor then being the determinant,

$$\begin{vmatrix} q_r(1) & q_r(2) & q_r(3) \\ q_b(1) & q_b(2) & q_b(3) \\ q_g(1) & q_g(2) & q_g(3) \end{vmatrix} \qquad (1.1)$$

This factor is unique, and all three-quark states which have this factor are referred to as "colour singlet". All baryonic states observed to date are colour singlets and we shall assume that this is necessarily so for all low-lying baryonic supermultiplets. Quite simple dynamical assumptions which ensure this situation have been specified by Lipkin.(3) With this assumption, it follows that the wavefunctions including space, spin and unitary-spin variables must have S symmetry, for the baryonic supermultiplets of interest to us here. Henceforth in these lectures, we shall assume this to be the case and we shall not mention the colour variable again, the wavefunction factor (1.1) being assumed always present.

The baryonic supermultiplets are then specified by their configuration ($\underline{N}$, LP), where $\underline{N}$ denotes the SU(6) representation. The space wavefunction then necessarily has the same permutation symmetry as $\underline{N}$. The parameters L and P denote the total orbital angular momentum and parity, given by

$$\underline{L} = \underline{\ell}_\rho + \underline{\ell}_\lambda, \qquad P = (-1)^{\ell_\rho + \ell_\lambda}, \qquad (1.2)$$

where $\ell_\rho$ and $\ell_\lambda$ denote the internal orbital angular momenta associated with the vectors $\underline{\rho}$ and $\underline{\lambda}$, given by

$$\underline{\rho} = (\underline{r}_1 - \underline{r}_2)/\sqrt{2}, \qquad \lambda = (\underline{r}_1 + \underline{r}_2 - 2\underline{r}_3)/\sqrt{6}, \qquad (1.3)$$

which specify the three-quark space configuration, as shown in Fig.1. The total spin of a particular sub-state of the supermultiplet is given by the vector addition

$$\underline{J} = \underline{L} + \underline{S}, \qquad (1.4)$$

where S is the spin associated with the SU(3) multiplet considered within the SU(6) multiplet $\underline{N}$. We shall find it convenient below to specify each supermultiplet by giving its value for L (and P), its symmetry type, and its degree of excitation. Each supermultiplet may then be specified by a point (L, E) on a plot with axes specifying L and excitation energy E, such as is shown in Fig.2.

The simplest type of plot possible is that shown in Fig.2, for the case of a diatomic molecule. Its energy E has the form

$$E = E(a) + L(L+1)\varepsilon(a), \tag{1.5}$$

where a specifies the vibrational state of the molecule, E(a) being the vibrational energy and $\varepsilon(a)$ giving the unit of rotational energy, the latter being inversely proportional to the moment of inertia of the molecule. The distinction made between the states L=even and L=odd on this Figure has a physical basis; for example, for the molecule $H_2$, the L=even states are those of the para-hydrogen molecule, whereas the L=odd states are those of ortho-hydrogen. The typical situation here is that $\varepsilon(a) \ll (E(a+1)-E(a))$ giving a sequence of rotational bands of states, each band being built on a different vibrational state; the L(L+1) dependence characteristic of these bands arises from the centrifugal term in the Hamiltonian, the vibrational motion being little affected by the rotation of the molecule.

For hadronic states, the indications are that the plot of L vs. excitation energy is linear for a rotational band when $(M^2-M_0^2)$ is used as abscissa in place of $(M-M_0)$, $M_0$ being the mass of the lowest supermultiplet. This is most readily understood in terms of harmonic quark-quark potentials, when a suitable wave equation is adopted for the three-quark system.[4] The linear dependence on L may be considered due to the softness of the harmonic potential; the moment of inertia is not constant but increases with increasing L. In the same way, the vibrational energies are low, comparable with the rotational energies; the unit of vibrational energy is $2\varepsilon$, where $\varepsilon$ denotes the unit of rotational energy. The mass formula is

$$M^2 = M_0^2 + n\varepsilon = M_0^2 + (L+2r)\varepsilon \tag{1.6}$$

where n is the number of quanta of excitation, and r gives the degree of vibrational excitation. The sequence of states with increasing L, for given r, constitutes a "Regge trajectory", in the language of the elementary particle physicist; it is well-known that, in general,[5] $\Delta L=2$ holds for successive states along a Regge trajectory.[6] The rotational sequence corresponding to r=0 is known as the leading Regge trajectory, that for r=1 is known as the first daughter trajectory, and the trajectory for r as the r-th daughter trajectory. If we assume that linear trajectories are the rule, then the existence of one state implies the existence of a Regge sequence of states on the Regge trajectory through

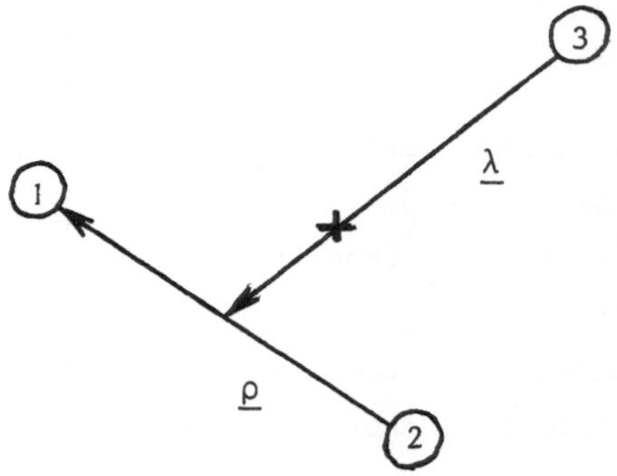


Fig.1. The co-ordinate system adopted here for a three-body system.

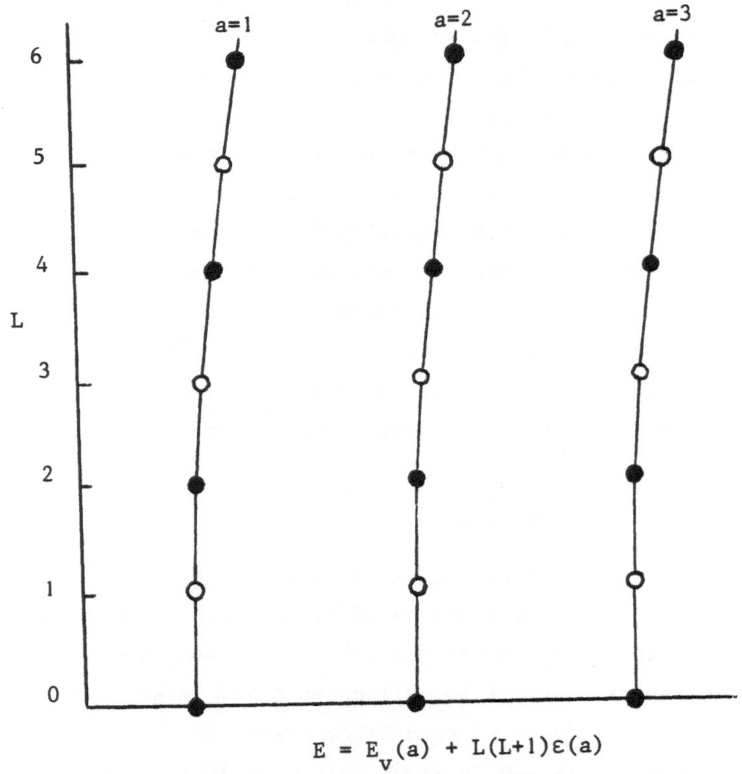


Fig.2. Pattern of Regge trajectories typical of a diatomic molecule. The parameter a characterizes the vibrational state of the molecule and L is its orbital angular momentum.

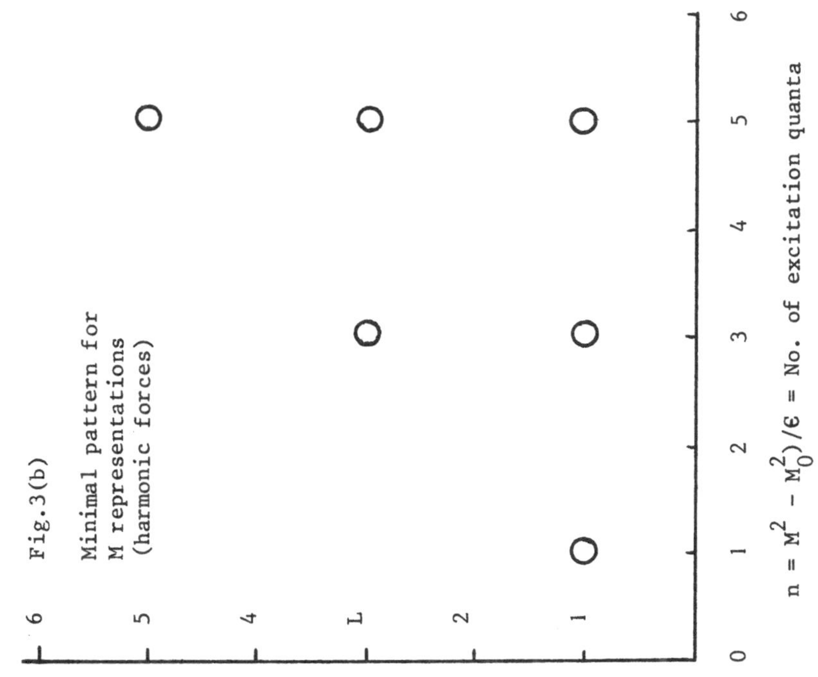
Fig.3(b)
Minimal pattern for
M representations
(harmonic forces)
$n = M^2 - M_0^2)/\epsilon$ = No. of excitation quanta

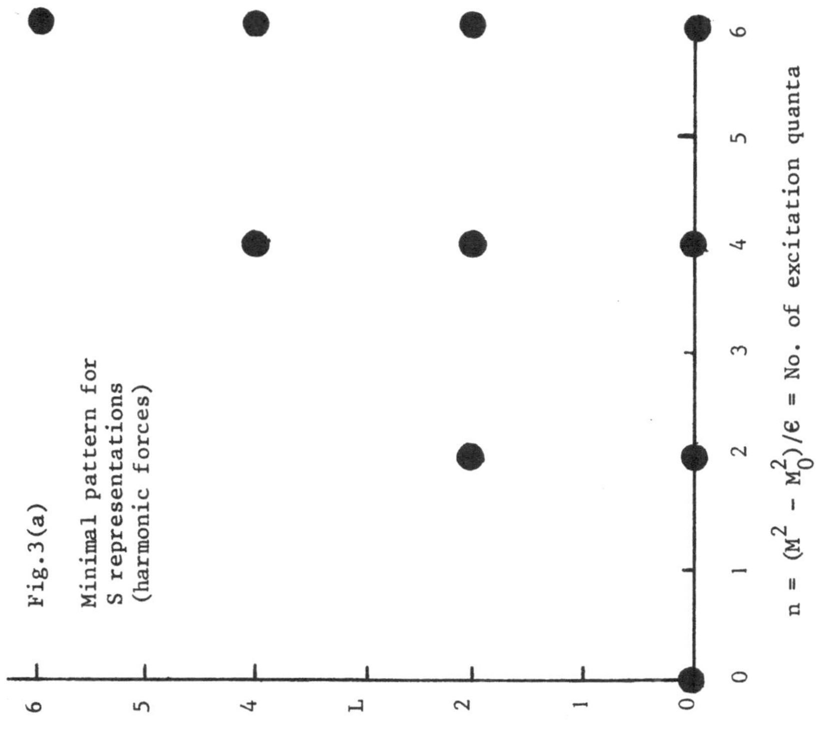
Fig.3(a)
Minimal pattern for
S representations
(harmonic forces)
$n = (M^2 - M_0^2)/\epsilon$ = No. of excitation quanta

it, together with the infinite sequence of daughter trajectories which lie below it and correspond to all the radial excitations of the states lying on the leading trajectory, according to rather fundamental principles.

The ground supermultiplet $(\underline{56}, 0+)_0$ for the baryonic states is well-known.[7] The corresponding point on the (L, n) plot is that at the bottom left of Fig.3(a), together with all the rotational and vibrational excitations derived from it. The first-excited supermultiplet is also well-established, with the structure $(\underline{70}, 1-)_1$. It has been entered similarly on Fig.3(b), together with all the radial and vibrational excitations derived from it. It is an empirical fact that these leading trajectories for the $\underline{56}$ and $\underline{70}$ representations are essentially coincident. This fact does not follow from any general principles and therefore represents rather significant dynamical information.[8] It happens to follow directly from the formula (1.6) derived from the harmonic-oscillator model for three-quark states. The patterns given for the representations $\underline{56}$ and $\underline{70}$ on Figs.3(a) and 3(b) have been plotted together on Fig.4(a), in order to indicate the minimal pattern of supermultiplets predicted on the basis of our knowledge of the $(\underline{56}, 0+)$ and $(\underline{70}, 1-)$ supermultiplets.

The (L, n) pattern of supermultiplets for the harmonic oscillator model has been considered in detail by Karl and Obryk[9] and their conclusions have been plotted on Fig.4(b). We see that this pattern of supermultiplets is far richer and more complicated than the minimal pattern on Fig.4(a). The number of supermultiplets for given degree of excitation n rises rapidly with increasing n, approximately like $n^3$. Even the leading trajectory soon becomes multiple. There is a $\underline{56}$ representation on it for every L value except L=1, and the leading $\underline{56}$ trajectory becomes double for $N \geq 6$, triple for $N \geq 12$, and so on.

The question is whether there is any evidence tc support this prediction of a rapidly increasing density of supermultiplets. Litchfield,[10] and later Cashmore et al.[11] have argued the case that the evidence available at present does not require the existence of any supermultiplets beyond those of the minimal pattern displayed on Fig.4(a). The $(\underline{56}, 2+)_2$ and $(\underline{56}, 0+)^*_2$ supermultiplets are now both established, in the sense that the nucleonic members are now known for every SU(3) multiplet contained in these supermultiplets, as follows:

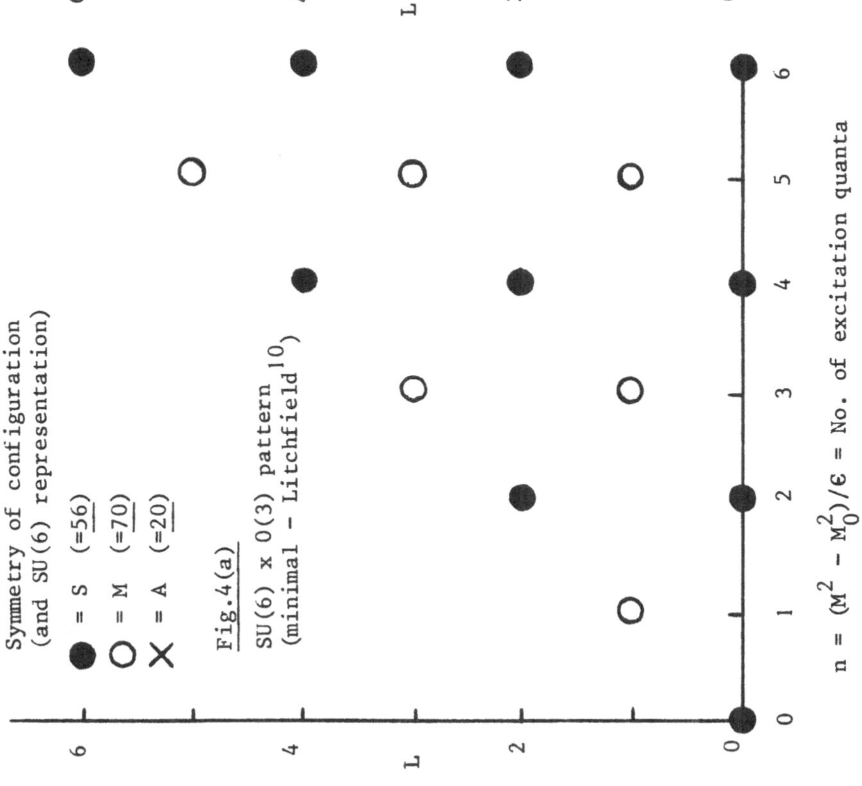


Fig.4(a)
SU(6) x O(3) pattern
(minimal - Litchfield[10])

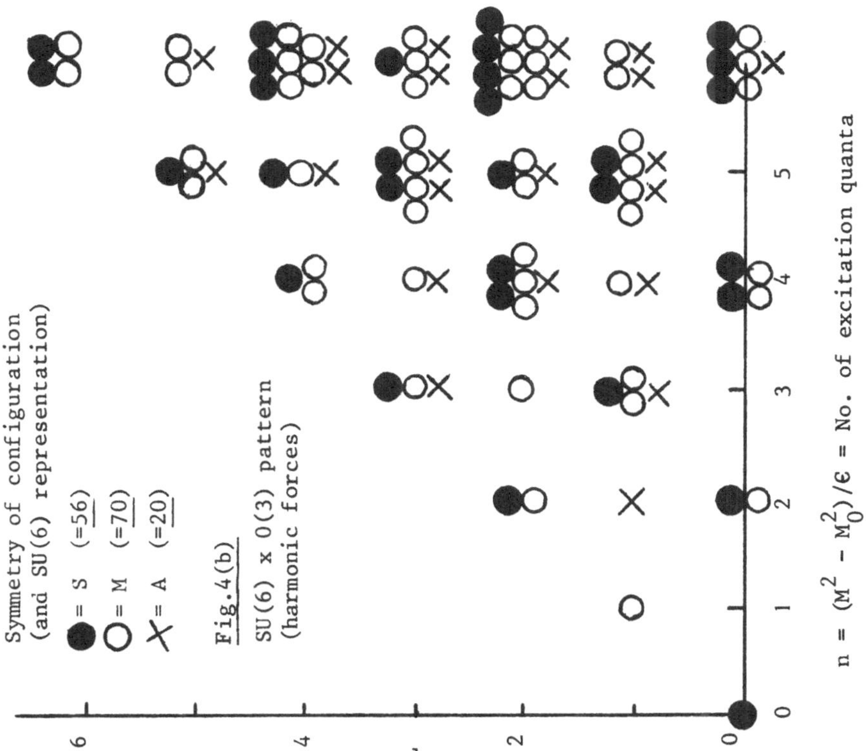


Fig.4(b)
SU(6) x O(3) pattern
(harmonic forces)

$$\underline{(56,\ 2+)_2}$$

$$^4D(\underline{10}) \begin{cases} \Delta F37(1925) \\ \Delta F35(1925) \\ \Delta P33(1900) \\ \Delta P31(1850) \end{cases} \quad ^2D(\underline{8}) \begin{cases} NF15(1680) \\ NP13(1730) \end{cases}$$

$$\underline{(56,\ 0+)_2^*}$$

$$^4S(\underline{10})\ \Delta P33(1690)$$
$$^2S(\underline{8})\ NP11(1450)$$

The controversy therefore concerns the rather limited number of states still remaining, most of which are not really well-established. Horgan[12] has assigned the following resonances to the $(\underline{70}, 2+)_2$ and $(\underline{70}, 0+)_2$ supermultiplets:

$(\underline{70}, 2+)_2$: NF17(1990), $\Lambda$F07(2020), $\Lambda$F05(2110), $\left[\Lambda F05(1815)\right]$, $\Lambda$P03(1860),
$\qquad \left[\Sigma F17(2030)\right]$

$(\underline{70}, 0+)_2$: NP11(1750), $\Lambda$P01(1770).

Of these states, the two in square brackets are well-established, but they are usually assigned to the $(\underline{56}, 2+)_2$ supermultiplet. In Horgan's analysis, their masses fit more naturally with the $(\underline{70}, 2+)_2$ supermultiplet; however, there are then no candidate resonances to fill the $\Lambda$F05 and $\Sigma$F17 gaps for the $(\underline{56}, 2+)_2$ supermultiplet, so that it is best to disregard these two assignments. Of the other states, only the NP11(1750) state has a three-star rating;[13] the states NF17(1990, $\Lambda$F07(2020), $\Lambda$P03(1860) and $\Lambda$P01(1770) have only a two-star rating. The decisive question is whether these states have an alternative interpretation possible, within the supermultiplets of Fig.4(a).

There are places empty for a $\Lambda$P03 resonance in the $(\underline{56}, 2+)_2$ supermultiplet, and for a $\Lambda$P01 resonance in the $(56, 0+)_2^*$, so that it is conceivable that the last two states just mentioned above might be fitted within established supermultiplets. As for the remaining three resonance states, Cashmore et al.[11] proposed to interpret them within the n=4 supermultiplets on Fig.3(a) as follows:

$(\underline{56}, 4+)_4$: NF17(1990), NH19(2250), $\Delta$H311(2400), $\Lambda$F07(2020), $\Lambda$H09(2350)
$(\underline{56}, 0+)_4^{**}$: NP11(1750).

Accepting these assignments, they made calculations of the partial widths expected for the various decay modes and concluded that these assignments were not ruled out by such data. However, it is apparent that these assignments do not fit comfortably with a mass formula analysis. The spin-orbit splittings for the two pairs of states NF17-NH19 and $\Lambda$F07-$\Lambda$H09 are of order 250-350 MeV, much larger than those found for any of the lower supermultiplets. For the nucleonic states of the $(\underline{56}, 2+)_2$ supermultiplet, for example, the over-all splittings are only ~75 MeV for the

$^4D(\underline{10})$ configuration and ~50 MeV for the $^2D(\underline{8})$ configuration. The largest spin-orbit splitting otherwise known is for the $(\underline{70}, 1-)_1$ pair of states ΛD03(1520)-ΛS01(1405), thus ~115 MeV.

The $(\underline{56}, 0+)_4^{**}$ interpretation for NP11(1750) would also place this n=4 supermultiplet very low in mass, not much above the $(\underline{70}, 1-)_1$ supermultiplet, in fact. However, the $(\underline{56}, 0+)_2^{*}$ supermultiplet lies even lower, so that such a possibility is not excluded. Cashmore et al.[11] argue that the photoproduction data indicates that NP11(1750) belongs to some $(\underline{56}, 0+)$ supermultiplet rather than to a $(\underline{70}, 0+)$ supermultiplet. The photoproduction parameter involved is $(A^P_{1-}/A^N_{1-})$ which would have the value +3 for a $(\underline{70}, 0+)$ assignment and +3/2 for a $(\underline{56}, 0+)$ assignment. The empirical value is +1.43±1.3, which really allows either possibility. With the $(\underline{56}, 0+)_4^{**}$ assignment, we also realise from Fig.3(a) that there should be another $(\underline{56}, 2+)_4^{*}$ supermultiplet lying somewhere between the supermultiplets $(\underline{56}, 0+)_4^{**}$ and $(\underline{56}, 4+)_4$, with mass value comparable to that for the $(\underline{56}, 2+)_2$ supermultiplet, but there are really no outstanding states (apart from the NP11(1750), ΛP03(1860) and ΛP01(1770) states under discussion here) which indicate the necessity for a second $(\underline{56}, 2+)$ supermultiplet in this mass region.

The empirical evidence for the existence of the $(\underline{70}, 2+)_2$ and $(\underline{70}, 0+)_2$ supermultiplets is rather slender, at the present time. The former depends primarily on the existence of the two states NF17(1990) and ΛF07(2020), both of which have only a two-star rating. The latter rests primarily on the interpretation of the one well-established state NP11(1750). It is clear that further experimental data on these resonances and a more thorough search for further nucleonic and hyperonic resonances in the mass region ~1850-2150 MeV could be rather fruitful at this stage.

## 2. The Harmonic-Oscillator Three-Quark Shell Model

(a) Introduction. In this chapter, we wish to discuss briefly how the complicated patterns of Fig.4(b) come about and on what elements of the model they depend. In terms of the co-ordinate vectors $\underline{\rho}$ and $\underline{\lambda}$ introduced earlier, the wavefunctions for a harmonic oscillator state with degree of excitation n have the general form

$$\psi_n = F_n(\underline{\rho}, \underline{\lambda})\mathrm{Exp}(-\tfrac{1}{2}\alpha^2(\rho+\lambda^2)) \qquad (2.1)$$

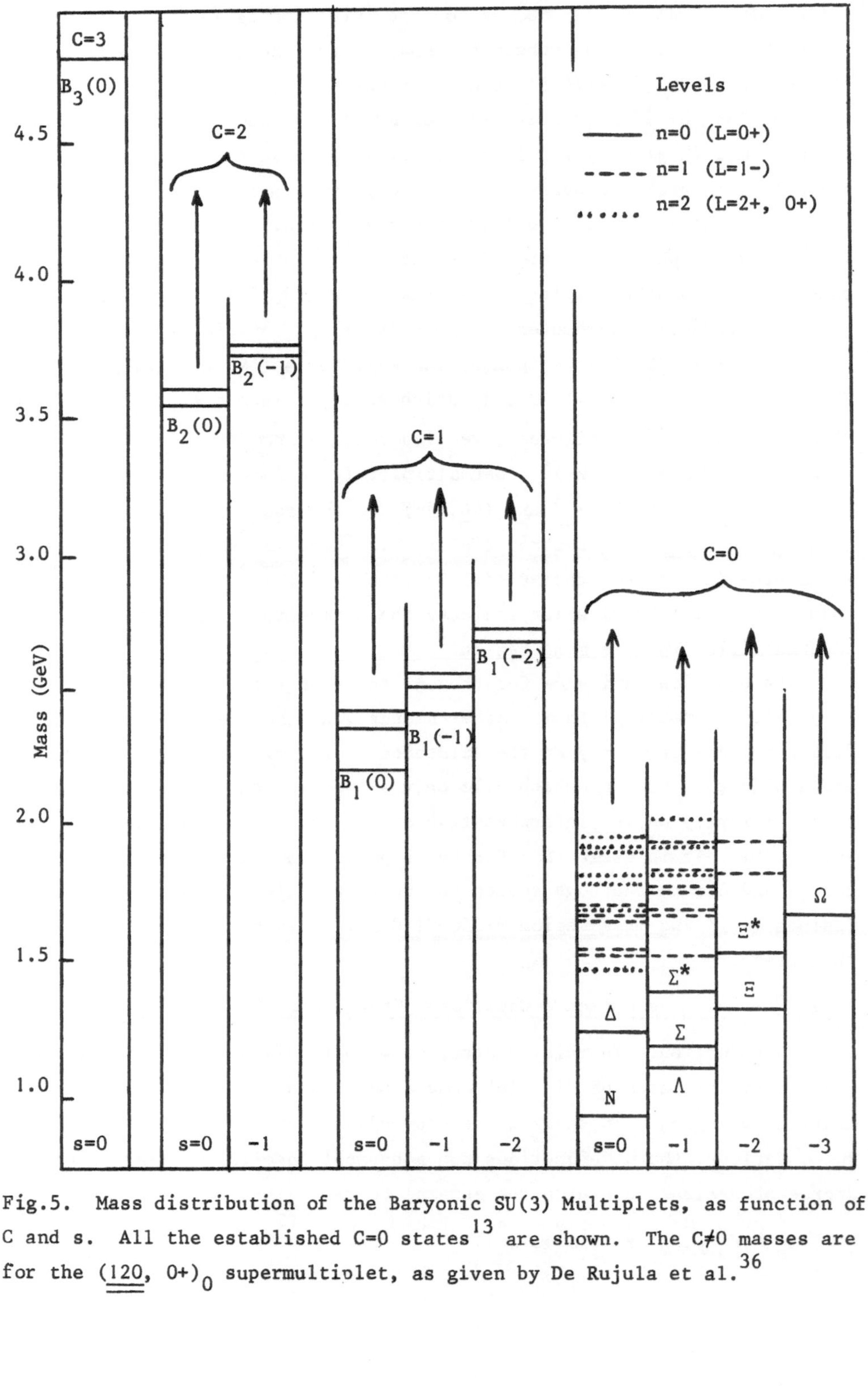


Fig.5. Mass distribution of the Baryonic SU(3) Multiplets, as function of C and s. All the established C=0 states[13] are shown. The C≠0 masses are for the ($\underline{\underline{120}}$, 0+)$_0$ supermultiplet, as given by De Rujula et al.[36]

where $F_n$ denotes a polynomial in the components $\rho_i$ and $\lambda_j$ of $\underline{\rho}$ and $\underline{\lambda}$, with powers up to a maximum of n. The wavefunction $\psi_n$ must be orthogonal to all of the wavefunctions $\psi_\nu$ for which $\nu<n$; in consequence, we need consider only those terms of $F_n$ which have the maximum degree n, since the terms of lower degree in $F_n$ are uniquely specified when the terms of maximum degree are given, through the orthogonality relations which ensure that the full wavefunction $\psi_n$ is orthogonal to all wavefunctions of lower degree. We denote the terms of degree n in $F_n$ by the n-th degree polynomial $\Phi_n(\underline{\rho}, \underline{\lambda})$.

Next, we consider the group $P_3$ of all permutations on the three labels 1, 2 and 3 attached to the three quarks. The co-ordinate vectors $(\underline{\lambda}, \underline{\rho})$ form the simplest non-trivial basis for an irreducible representation of $P_3$; they give a mixed representation, as we see for the permutations P(12) and P(13)

$$P(12)\begin{pmatrix}\lambda_i\\ \rho_i\end{pmatrix} = \begin{pmatrix}1 & 0\\ 0 & -1\end{pmatrix}\begin{pmatrix}\lambda_i\\ \rho_i\end{pmatrix}, \qquad P(13)\begin{pmatrix}\lambda_i\\ \rho_i\end{pmatrix} = \begin{pmatrix}-\frac{1}{2} & -\frac{1}{2}\sqrt{3}\\ -\frac{1}{2}\sqrt{3} & \frac{1}{2}\end{pmatrix}\begin{pmatrix}\lambda_i\\ \rho_i\end{pmatrix} \tag{2.2}$$

from which all the permutations of the group $P_3$ can be built up. From this basis $(\lambda_i, \rho_i)$ all the polynomials which give irreducible representations of $P_3$ can be built up by successive multiplications and reductions. For this purpose, we must now specify the multiplication tables for the reduction of the product of two irreducible representations, as follows:

$$\text{(i) } M \otimes M = S \oplus M \oplus A, \text{ (ii) } M \otimes S = M, \text{ (iii) } M \otimes A = M, \tag{2.3}$$

together with the trivial products

$$S \otimes S = S, \quad S \otimes A = A, \quad A \otimes A = S. \tag{2.4}$$

We shall need to have explicit expressions for the irreducible representations obtained in the reduction of the first product in line (2.3). If the bases for these irreducible M representations are taken to be $(\Lambda, R)$ and $(\Lambda', R')$ respectively, the irreducible representations obtained from the reduction (2.3(i)) have the following base states:

$$\text{(a) S: } (\Lambda\Lambda' + RR'), \quad \text{(b) M: } \begin{pmatrix}RR' - \Lambda\Lambda'\\ R\Lambda' + R'\Lambda\end{pmatrix}, \quad \text{(c) A: } (\Lambda R' - \Lambda' R) \tag{2.5}$$

Next, we shall consider three subsets of states, which have a particular simplicity.

(b) <u>The Leading Trajectory and its Daughter Trajectory.</u> The states on the leading trajectory have maximum L for given n, namely L=n. For these states, we need consider only the substates with magnetic quantum number $m_L$=L=n. The corresponding homogeneous polynomial $\Phi_n$ of degree n is then made up of $\lambda_+$ and $\rho_+$ only, since each co-ordinate factor in each term of $\Phi_n$ must contribute m=+1 to the net sum $m_L$=n.

We start with the polynomial representation of degree n=1, namely $\mathcal{M}_1 = (\lambda_+, \rho_+)$. From this, by multiplying $\mathcal{M}_1$ by itself, and using the reduction tables (2.5), we obtain the further irreducible polynomial representations now given:

$$\text{(i)}\ \mathcal{S}_2 = (\lambda_+^2 + \rho_+^2), \quad \text{(ii)}\ \mathcal{M}_2 = \begin{pmatrix} \rho_+^2 - \lambda_+^2 \\ 2\rho_+\lambda_+ \end{pmatrix}, \quad \text{(iii)}\ \mathcal{A}_2 \equiv 0 \tag{2.6}$$

We repeat this process by considering $\mathcal{M}_1 \otimes \mathcal{M}_2$, with the results:

$$\text{(i)}\ \mathcal{S}_3 = (3\lambda_+\rho_+^2 - \lambda_+^3), \quad \text{(ii)}\ \mathcal{M}_3 \equiv \mathcal{S}_2\mathcal{M}_1, \quad \text{(iii)}\ \mathcal{A}_3 = (3\rho_+\lambda_+^2 - \rho_+^3). \tag{2.7}$$

Again, for the product $\mathcal{M}_2 \otimes \mathcal{M}_2$, we then have

$$\text{(i) S:}\ \mathcal{S}_2^2, \quad \text{(ii) M:}\ \mathcal{M}_4 = \begin{pmatrix} 4\rho_+^2\lambda_+^2 - (\rho_+^2-\lambda_+^2)^2 \\ 4\rho_+\lambda_+(\rho_+^2-\lambda_+^2) \end{pmatrix}, \quad \text{(iii) A: zero.} \tag{2.8}$$

At this point, the process comes to a halt; there are no further irreducible polynomials which are not simply products of some power of the symmetrical polynomials $\mathcal{S}_2$ and $\mathcal{S}_3$ with <u>at most one factor</u> of $\mathcal{M}_1$, $\mathcal{M}_2$, $\mathcal{M}_4$ or $\mathcal{A}_3$. To demonstrate this, we consider the reduction of the further products, the right-hand side representing in turn the S, M and A terms (with the correct algebraic coefficients) resulting:

$$\mathcal{M}_2 \otimes \mathcal{M}_3 = \mathcal{S}_2\mathcal{S}_3 \oplus (\mathcal{S}_3\mathcal{M}_2 + \mathcal{A}_3\overline{\mathcal{M}}_2) \oplus \mathcal{S}_2\mathcal{A}_3, \tag{2.9a}$$
$$\mathcal{M}_2 \otimes \mathcal{M}_4 = (\mathcal{S}_2^3 + \mathcal{S}_3^2) \oplus \mathcal{S}_2^2\mathcal{M}_2 \oplus (-2\mathcal{S}_3\mathcal{A}_3), \tag{2.9b}$$
$$\mathcal{M}_4 \otimes \mathcal{M}_4 = \mathcal{S}_2^4 \oplus (2(2\mathcal{S}_3^2 - \mathcal{S}_2^3)\mathcal{M}_2 - \mathcal{S}_2^2\mathcal{M}_4) \oplus \text{zero}, \tag{2.9c}$$
$$\mathcal{A}_3 \otimes \mathcal{A}_3 = (\mathcal{S}_2^3 - \mathcal{S}_3^2) \oplus \text{zero} \oplus \text{zero}, \tag{2.9d}$$

where $\overline{\mathcal{M}}$ denotes $(-R, \Lambda)$ when $\mathcal{M}$ denotes $(\Lambda, R)$. These relations show that the set of homogeneous polynomials is then closed, and that

(i) $\mathcal{S}_2$ and $\mathcal{S}_3$ are the only independent polynomials with S symmetry,

(ii) $\mathcal{M}_1$, $\mathcal{M}_2$ and $\mathcal{M}_4$ are the only polynomial pairs forming a basis for M symmetry,

(iii) $\mathcal{A}_3$ is the only polynomial with A symmetry.

We also note explicitly that $\mathcal{S}_3^{\,2} \neq \mathcal{S}_3^{\,2}$, as is apparent from their forms, given in Eqs.(2.6) and (2.7).

The states on the leading trajectory now correspond to the homogeneous polynomials which can be formed as products of some power of $\mathcal{S}_2$ and $\mathcal{S}_3$, together with a factor 1 for S symmetry, a factor $\mathcal{M}_1$, $\mathcal{M}_2$ or $\mathcal{M}_4$ for M symmetry, or a factor $\mathcal{A}_3$ for A symmetry. Let us consider the sequences in turn:

(i) S symmetry. The Regge sequence of states based on the ground state is represented by the sequence of polynomials

$$1, \mathcal{S}_2, \mathcal{S}_2^{\,2}, \mathcal{S}_2^{\,3}, \mathcal{S}_2^{\,4}, \mathcal{S}_2^{\,5}, \ldots, \tag{2.10}$$

corresponding to the usual intervals $\Delta L=2$ along a trajectory. However, because of the three-particle structure of these states, we see that there is also a state with L=3, corresponding to the function $\mathcal{S}_3$, which commences a new sequence

$$\mathcal{S}_3, \mathcal{S}_3\mathcal{S}_2, \mathcal{S}_3\mathcal{S}_2^{\,2}, \mathcal{S}_3\mathcal{S}_2^{\,3}, \mathcal{S}_3\mathcal{S}_2^{\,4}, \ldots \tag{2.11}$$

giving a second Regge sequence on the leading trajectory. More generally we see that a new Regge sequence ($\Delta L=2$) commences at L=3s, for every integer s, consisting of the states corresponding to $\mathcal{S}_3^{\,s}$, $\mathcal{S}_3^{\,s}\mathcal{S}_2$, $\mathcal{S}_3^{\,s}\mathcal{S}_2^{\,2}$, $\mathcal{S}_2^{\,s}\mathcal{S}_2^{\,3}$, ... For given n, the S states first become double at n=6, triple at n=12, and so on.

(ii) M symmetry. The established $(\underline{70}, 1-)_1$ supermultiplet corresponds to $\mathcal{M}_1$, and the Regge sequence based on this state corresponds to $\mathcal{M}_1$, $\mathcal{M}_1\mathcal{S}_2$, $\mathcal{M}_1\mathcal{S}_2^{\,2}$... However, a Regge sequence of M symmetry supermultiplets also begins from each polynomial $\mathcal{M}_1\mathcal{S}_3^{\,s}$, thus from L=(3s+1), for every integer s. Similar sequences are based on the polynomials $\mathcal{M}_2$ and $\mathcal{M}_4$. To each homogeneous polynomial

$$\mathcal{M}_\alpha \mathcal{S}_3^{\,s} \mathcal{S}_2^{\,\sigma} \tag{2.12}$$

for $\alpha$ = 1, 2 or 4, and s and $\sigma$ integral, there corresponds a state on the leading trajectory with internal orbital angular momentum $L = (\alpha+3s+2\sigma)$. These states are arranged as Regge sequences for $\sigma$ = 0, 1, 2, 3 ..., starting from an M supermultiplet with $n=L=(\alpha+3s)$.

(iii) A symmetry. The first A supermultiplet on the leading trajectory is for L=3, and there is a series of these with $\Delta L=3$, corresponding to the polynomials $\mathcal{A}_3\mathcal{S}_3^{\,s}$. There is a normal Regge sequence starting from each of these states, i.e. starting from a new A supermultiplet with L=3(s+1), corresponding to the sequence of polynomials

$$\mathcal{A}_3\mathcal{S}_3^{\,s}\mathcal{S}_2^{\,\sigma} \qquad (2.13)$$

for $\sigma$ = 0, 1, 2 ...

The additional complexity of these patterns (shown in Fig.4(b)) over that characteristic of a diatomic system (as displayed in Fig.4(a) for harmonic forces) arises in two ways:

(a) the three-particle character of the system allows the existence of a number of basic structures, corresponding to different permutation symmetries for the internal space wavefunction. For each of these basic patterns, given by polynomials 1, $\mathcal{M}_1$, $\mathcal{M}_2$, $\mathcal{M}_4$, and $\mathcal{A}_3$, there is naturally a Regge sequence which corresponds in our algebraic discussion to the multiplication of these polynomials by $\mathcal{S}_2^{\,\sigma}$, for $\sigma$ = 1, 2, ... This multiplies the number of states by a factor 5 over those for the diatomic case.

(b) the three-particle character of the system allows the existence of a second polynomial $\mathcal{S}_3$ with S symmetry. This leads to a second sequence of states, corresponding to the multiplication of each of the polynomials just referred to under (a) by a factor $\mathcal{S}_3^{\,s}$, for integers s = 1, 2, 3 ... It is this second feature which gives rise to the vastly greater number of supermultiplets in the three-quark model for the degree of excitation n.

At this point, it is convenient to consider the first daughter trajectory, with L=(n-1). Here the wavefunctions have the same dimensionality n, but the state of maximum magnetic quantum number has $m_L$=(n-1). These states are constructed by contracting the product $\rho_i\lambda_j$ to form $(\underline{\lambda}\times\underline{\rho})_+$ which has angular momentum 1 and $m_L$=+1. All the remaining (n-2) factors must now have $m_L$=+1 in order that, together with $(\underline{\lambda}\times\underline{\rho})_+$, they reach the required value $m_L$=(n-1). The factor $(\underline{\lambda}\times\underline{\rho})_+$ is itself antisymmetric, and must be present for each term in the polynomial. Hence the polynomial it multiplies must be such as to give orbital angular momentum L=(n-2) for a homogeneous polynomial with degree (n-2) and to correspond to a definite permutation symmetry S, M or A, according as we require A, M, or S symmetry for the daughter state with L=(n-1). However these last polynomials are just those which we have discussed above for the states on the leading trajectory. In other words, the wavefunctions for the first daughter trajectory are obtained simply by multiplying those for the leading trajectory by the factor $(\underline{\lambda}\times\underline{\rho})_+$; the symmetry of the first daughter state is then changed from that of the leading trajectory state by the additional factor A. The correspondence between the leading and first daughter trajectories is indicated by the light arrowed lines on Fig.4(b).

(c) Radially-Excited States with L=0. Here the wavefunction for degree of excitation n is again completely characterised by a homogeneous polynomial of degree n in the components $\rho_i$ and $\lambda_j$ of the vectors $\underline{\rho}$ and $\underline{\lambda}$, but now the polynomial must be a scalar quantity. Since the only available scalars are $\rho^2$, $\underline{\rho}.\underline{\lambda}$, and $\lambda^2$, it is clear that such a wavefunction exists only for n=even.

We start with the M representation given by the co-ordinate vectors, $M_{1\alpha} = (\lambda_\alpha, \rho_\alpha)$. From the product $M_{1\alpha}M_{1\beta}$ we can form combinations with S and M symmetry,

$$(S_2)_{\alpha\beta} = (\lambda_\alpha\lambda_\beta+\rho_\alpha\rho_\beta), \quad (M_2)_{\alpha\beta} = \begin{pmatrix}\rho_\alpha\rho_\beta-\lambda_\alpha\lambda_\beta\\ \rho_\alpha\lambda_\beta+\lambda_\alpha\rho_\beta\end{pmatrix} \tag{2.14}$$

from which, by contracting on $\alpha\beta$, we obtain the scalars

$$S_2 = (\lambda^2+\rho^2), \qquad M_2 = \begin{pmatrix}\rho^2-\lambda^2\\ 2\underline{\rho}.\underline{\lambda}\end{pmatrix} \tag{2.15}$$

From the product $M_2 \otimes M_2$, we generate two distinct fourth-degree scalars $S_4$ and $M_4$, given by

$$S_4 = (\rho^2-\lambda^2)^2 + 4(\underline{\rho}.\underline{\lambda})^2, \qquad M_4 = \begin{pmatrix}(\rho^2-\lambda^2)^2 - 4(\underline{\rho}.\underline{\lambda})^2\\ 4\underline{\lambda}.\underline{\rho}(\rho^2-\lambda^2)\end{pmatrix} \tag{2.16}$$

The product $M_2 \otimes M_4$ leads to the reduction

$$M_2 \otimes M_4 = S_6 \oplus S_4M_2 \oplus A_6 \tag{2.17}$$

where

$$S_6 = 12(\underline{\rho}.\underline{\lambda})^2(\rho^2-\lambda^2) - (\rho^2-\lambda^2)^3, \quad A_6 = 6\underline{\rho}.\underline{\lambda}(\rho^2-\lambda^2)^2 - 8(\underline{\rho}.\underline{\lambda})^3. \tag{2.18}$$

All other scalar polynomials giving irreducible representations of the permutation group may be expressed in terms of these; for example, $A_6^2$ has S symmetry and may be written in the form $(S_6^2-S_4^3)$. The proof that this is the case has been given by Karl and Obryk [9] (and by others [14]) by an enumeration of linearly independent forms, sufficient in number to agree with the number of representations of each type calculated by the method of characters.

For each type of symmetry, the wavefunctions are generated by multiplying the appropriate base function $\chi_\lambda$, which is $\chi_0=1$ for S symmetry, $\chi_2=M_2$ or $\chi_4=M_4$ for M symmetry, and $\chi_6=A_6$ for A symmetry, by integral powers of $S_2$, $S_4$ and $S_6$, to give the sequences

$$\Phi_{n,\lambda}(L=0) = \chi_\lambda(S_2)^\alpha(S_4)^\beta(S_6)^\gamma, \text{ where } n=\lambda+2\alpha+4\beta+6\gamma. \tag{2.19}$$

The number of partitions $[\alpha, \beta, \gamma]$ such that $(\gamma+2\beta+3\gamma) = (n-\lambda)/2$ increases rapidly as n increases for given $\lambda$, as is clearly apparent on the L=0 axis of Fig.4(b).

(d) <u>All Other Supermultiplets</u>. The polynomials $\Phi_n(L)$ for the general case can be constructed from the above polynomials, according to plausible rules which have been proved by Karl and Obryk.[9] These are as follows:

(i) (n-L)=even. If X and Y denote the symmetry classes (S, M or A) considered, then we take the $m_L=L$ polynomial $\Phi_{L,X}(L)$ for a state on the leading trajectory and multiply it by the scalar function $\Phi_{(n-L),Y}(0)$. This gives the required wavefunction for each of the symmetry classes contained in the product $X \otimes Y$.

For example, consider n=4, L=2. There are two leading trajectory states for L=2, with S and M symmetry, and there are two n=2, L=0 excitations, with S and M symmetry. The combinations $S \otimes S$, $S \otimes M$ twice, and $M \otimes M$ generate two S, three M and one A state, as is indicated for the point (n, L) = (4, 2) on Fig.4(b).

(ii) (n-L)=odd. Since radial excitations occur only for an even degree of excitation, it is necessary to consider radial excitations corresponding to the even degree (n-L-1). The remaining polynomial factors then correspond to states on the first daughter trajectory which was already discussed above in Sec.2(b). The required polynomials with m=(L-1) are then obtained by forming the product of each $\Phi_{L+1,X}(L)$ from the first daughter trajectory with each scalar function $\Phi_{(n-L-1),Y}(0)$, for each of the symmetry classes contained in the product $X \otimes Y$.

For example, consider n=6, L=3. The first daughter trajectory for L=3 has two states, with M and A symmetry (which may be derived, as discussed in Sec.2(b), from the two leading trajectory states for L=2), and there are two n=2, L=0 excitations, with S and M symmetry. The combinations $S \otimes A$, $S \otimes M$, $M \otimes A$ and $M \otimes M$ generate one S, three M and two A supermultiplets, as indicated on Fig.4(b).

In this way, the complete plot of Fig.4(b) may be built up systematically, for all n and L. The procedure of construction leads to a unique (albeit rather complicated) specification of every supermultiplet on this plot.[15-17]

(e) Modified Models. Here we mention briefly some distortions of the harmonic oscillator model, which have been proposed to reduce the high density of supermultiplets predicted for the mass region above about 2 GeV, a region already much studied experimentally.

(i) quark-diquark models. Lichtenberg[18] and Ono[19] have made the *ad hoc* assumption that two quarks have an especially strong interaction in the $\ell_{12}=0$ state, which limits the pair to the $\ell_{12}=0$ configuration. However, the exchange of a quark between the q and (qq) systems is permitted, giving rise essentially to an exchange interaction between them; the proper three-quark permutation symmetry is maintained in this model. Not surprisingly, the result is to give (L, $M^2$) plots with patterns appropriate to a diatomic system with exchange forces, of the types given in Figs.3(a) and 3(b). Capps[20] has discussed some special assumptions concerning the colour characteristics of the quark-quark force and which lead to forces which could yield a quark-diquark structure for the three-quark system.

(ii) Eguchi[21] has discussed some general considerations concerning the energetics of the various configurations ($\ell_{12}$, $\ell_3$) for a three-quark system, in the high excitation limit, in consequence of a linear relation between $E^2$ and $\ell$ for any two-particle system. He considers only "stretched systems", such that $L=\ell_{12}+\ell_3$, concluding that the two extreme configurations ($\ell_{12}$, $\ell_3$) = (0, L) and (L, 0) are energetically favoured. The spectrum is then again like that for a diatomic system; all of the other states for the three-quark system still exist but are then pushed up to high mass values.

(iii) Mitra[22, 23] has proposed that the quark-quark force should have a strong space-exchange character, of the form $(1-\varepsilon+\varepsilon P_x)$ with $\varepsilon=1/2$. The three-quark states with S symmetry are unaffected by this modification since $\ell_{12}$=even always holds for the q-q interactions in this case. However, the net attraction in the three-quark states with M symmetry is strongly reduced, while there is no attraction at all in the states with A symmetry, for $\varepsilon\geq 1/2$. Although the details are not all clear, it is apparent that the lowest (70, 1−) supermultiplet would be pushed very high in mass value relative to the $(\underline{56}, 0+)_0$ and $(\underline{56}, 2+)_2$ supermultiplets whereas the 56 supermultiplets with L≠even or P≠+, such as the $(\underline{56}, 3-)_3$ supermultiplet will be left unaffected in mass.

## 3. The New Spectroscopy

In November 1974, there emerged dramatic evidence for a new degree of freedom in hadronic physics, with the discovery of the remarkable ψ/J vector particle of mass 3.1 GeV.[24] This was quickly followed by the discovery of a second vector particle ψ(3.7), and later by the observation of a broad bump in the total cross section $\sigma(e^+e^- \rightarrow \text{hadrons})$ in the region of mass 4.2 GeV.[24] This latter structure has recently been resolved[25] into two resonance peaks, one at 3.95 GeV with width about 50 MeV, and the other at 4.4 GeV with width about 35 MeV, together with a broad hump (which may yet be resolved further) lying between them, at about 4.1 GeV, with width of order 150 MeV. Each of these peaks corresponds to a further vector meson of the ψ family. Besides these vector states, we now know that further states exist for this ψ meson family, with JP≠(1−), which cannot be excited directly in $e^+e^-$ collisions.[26] In any case, from the JP=(1−) members alone, it is clear that this ψ meson family must have a rich spectroscopy.

The precise nature of these ψ particles is not yet known. The simplest possibility not yet excluded is to suppose that there is a fourth quark, in addition to the quarks (u, d, s) with which we have been interpreting the mesonic and baryonic spectroscopy so admirably summarized in the 1974 Review of Particle Properties.[13] This is a possibility proposed long ago by Bjorken and Glashow,[27] who proposed the name "Charm" for the attribute of this fourth quark, so we shall denote this quark by c and its attribute by C, although it is not necessarily the case that this quark c should have all the properties envisaged for it by these authors, or by Glashow et al.[28] more recently. Of course, there also exist more complicated possibilities, but it will be sufficient for our purpose here to confine attention to the simple case of one additional quark.

This quark c has value +1 for the quantum number C. Its introduction closely parallels the introduction of the strange quark, with value −1 for the strangeness quantum number s. Thus, for any hadronic system, the value of C is given by

$$C = (\text{number of c quarks} - \text{number of } \bar{c} \text{ quarks}), \tag{3.1}$$

just as s is given by

$$s = (\text{number of } \bar{s} \text{ quarks} - \text{number of s quarks}). \tag{3.2}$$

In this respect, we can expect ψ physics to parallel strange-particle physics, apart from some important practical differences to be discussed later. For example, the ψ particles themselves are interpreted as $^3S_1$ states of the $\bar{c}$-c system, with radial excitations, just as the φ(1020) meson is interpreted as the lowest $^3S_1$ state of the $\bar{s}$-s system and ω(785) as the I=0 component of the lowest $^3S_1$ states for the $\bar{u}$-u and $\bar{d}$-d systems.

Even the existence of a new additive quantum number has not been sufficient to explain the most striking property of ψ/J(3.1) and ψ(3.7), namely, their lifetimes of order $10^{-20}$sec, remarkably long when measured on the time-scale (unit ~$10^{-23}$sec) appropriate to hadronic phenomena. To account for this, it has been necessary to appeal to the "Zweig rule", whose operation is best illustrated by the relatively long lifetime (~$10^{-22}$sec) known for the φ(1020) meson.[29] To emphasize the bizarre nature of our present situation, we may add here that, since the established ψ particles can all be produced by the reaction

$$e^+ + e^- \rightarrow "\gamma" \rightarrow \psi, \tag{3.3}$$

they necessarily have quantum numbers compatible with those of the photon; in particular, they must have the value C=0 for this new quantum number. In other words, no hadronic state which has a non-zero value for this new quantum number C is yet known.

When there is only one additional quark c, it is necessarily a unitary singlet, since SU(3) symmetry is concerned only with unitary transformations in the space of (u, d, s) quarks. The charge of the c quark can have any value differing from +2/3 by an integer, in principle, but we shall follow the usual assumption,[30] that $Q_c = +2/3$, just as is the case for the u quark. The charge of any hadronic state is then given by

$$Q = I_3 + (B + s + C)/2. \tag{3.4}$$

We recall that B = 1/3 for the quark c, as for the quarks (u, d, s).

Mesonic states with C≠0 will result from the combination of quark c with the ($\bar{u}$, $\bar{d}$, $\bar{s}$) antiquarks. These consist of an isospin doublet D and a strange isospin singlet S, where we have adopted the names introduced by Gaillard et al.[31],

$$(C, s) = (+1, 0): \qquad D = (D^+, D^0) = (c\bar{d}, c\bar{u}), \qquad (3.5a)$$

$$(C, s) = (+1, +1): \qquad S = (c\bar{s}). \qquad (3.5b)$$

Their antiparticle states are denoted by $\bar{D} = (D^0, \bar{D}^-)$ and $\bar{S}^-$. The relationship of their masses with those for the normal states $\omega = (\bar{u}u + \bar{d}d)/\sqrt{2}$, $\phi = (\bar{s}s)$, $K^* = (\bar{s}u, \bar{s}d)$, and $\rho = (\rho^+, \rho^0, \rho^-)$ with $\rho^+ = (u\bar{d})$, etc., depends first of all on the mass relationship between the c quark and the (u, d, s) quarks, and on the relationship of the $\bar{c}$-c and $\bar{c}$-(u, d, s) forces with those of ($\bar{u}$, $\bar{d}$, $\bar{s}$) with (u, d, s). In the literature,[31] much emphasis has been placed on the possibility that these forces obey an SU(4) symmetry, i.e. invariance with respect to all unitary transformations with modulus unity in the space of all four quarks (u, d, s, c). In this case, the mass separations between the vector states $\omega$, $\rho$, $K^*$, $\phi$, D, S, and $\psi$ are due primarily to "mass differences" between the four quarks, although SU(4)-breaking $\bar{q}$-q forces can also contribute. The simple model, in which each quark contributes a "one-body operator" or "quark mass term" to the net $(\text{mass})^2$ for the state, works extremely well for the subset ($\rho$, $\omega$, $K^*$, $\phi$) of vector mesons. If we follow the same prescription for the new vector mesons, then we have the result

$$2m(D)^2 = m(\omega)^2 + m(\psi)^2 \longrightarrow m(D) = 2.26 \text{ GeV}, \qquad (3.6a)$$

$$2m(S)^2 = m(\phi)^2 + m(\psi)^2 \longrightarrow m(S) = 2.30 \text{ GeV}. \qquad (3.6b)$$

The use of a linear mass formula would give slightly lower values, $m(D) = 1.94$ GeV and $m(S) = 2.06$ GeV. Since the $\psi$ mass is so disparate with the $\omega$ and $\phi$ masses, it is far from obvious that this simple calculation should be correct, even if its physical assumptions were valid; the observation of $C\neq0$ mesons will give us illuminating guidance in this respect. In the interim, the estimates (3.6) give a useful orientation. It is worth remarking that this model requires the existence of corresponding pseudoscalar states $D_p$ and $S_p$, with $(c\bar{q})$ configuration $^1S_0$, their masses being given by

$$m(D_p)^2 - m(D)^2 = m(S_p)^2 - m(S)^2 = m(\pi)^2 - m(\rho)^2 = m(K)^2 - m(K^*)^2, \qquad (3.7)$$

assuming the quadratic mass formula (as is required for the correctness of the last equality in (3.7)); with the Eqs.(3.6), Eq.(3.7) then leads to the meson mass values $m(D_p) = 2.13$ GeV and $m(S_p) = 2.16$ GeV. Indeed, it is now generally assumed that the masses of the D and S mesons are in the vicinity of 2 GeV and that the onset of $e^+e^- \to \bar{D}D$ pair creation is at least partly responsible for the rise of the ratio $R = \sigma(e^+e^- \to \text{Had.})/\sigma(e^+e^- \to \mu^+\mu^-)$ between about 3.5 GeV and 5.0 GeV.[32]

The C≠0 baryonic states may be discussed on the basis of the quark shell model. As for the (u, d, s) quarks, the c quark must be endowed with colour, transforming according to the SU(3)' group in this colour space, and all low-lying three-quark states are expected to be colour singlets. Furthermore, we shall assume that the orbitals for the c quark are the same as those for the (u, d, s) quarks. Then, the three-quark states which belong to the $(\underline{56}, 0+)_0$ supermultiplet and which have C=0 and represent the known octet and decuplet baryons, are supplemented by the following states:

(i) the C=+1 states cqq. Their SU(3) character is given by $3 \otimes 3 = 6 \oplus \bar{3}$. The 6 states have even permutation symmetry for the SU(3) variables, and this requires parallel spin, $S_{qq}=1$, since we are concerned with the ground configuration $(\ell_c, \ell_{qq})=(0, 0)$. The $\bar{3}$ states have odd permutation symmetry for the SU(3) variables, which requires $S_{qq}=0$. The 6 states thus have J=3/2 or J=1/2, whereas the $\bar{3}$ states necessarily have J=1/2. The J=3/2 states correspond to the C=0 baryon decuplet, whereas the J=1/2 states correspond to the C=0 baryon octet.

(ii) the C=2 states ccq. These form a 3 representation. The cc symmetry requires $S_{cc}=1$, which allows both J=3/2 and J=1/2. Again, the J=3/2 states correspond to the C=0 decuplet, while the J=1/2 states correspond to the C=0 octet.

(iii) the C=3 states ccc. The cc symmetries require all three spins to be parallel, giving J=3/2, corresponding to the C=0 decuplet.

We see two distinct patterns of states, those with J=3/2 and corresponding to the known C=0 decuplet, and those with J=1/2 and corresponding to the known C=0 octet.

If we go to the limit of SU(4) symmetry, these J=3/2 states are the substates of a 20-dimensional representation with even permutation symmetry, which we denote by $\underset{\sim}{20}_S$, whereas these J=1/2 states are the substates of a 20-dimensional representation with mixed permutation symmetry, which we denote by $\underset{\sim}{20}_M$. Thus

$$\underset{\sim}{20}_S = (10)_0 + (6)_1 + (3)_2 + (1)_3, \quad \underset{\sim}{20}_M = (8)_0 + (6)_1 + (\bar{3})_1 + (3)_2 \tag{3.8}$$

where the notation $\underset{\sim}{N}$ denotes an SU(4) representation, its suffix giving the permutation symmetry for a three-quark state, and the $(\alpha)_C$ denote (SU(3), C) representations. There is one further SU(4) representation appropriate to the qqq system, namely $\bar{4}_A = (1)_0 + (\bar{3})_1$.

To a first approximation, the mass spectrum for baryonic states results from the "quark mass differences" (= the one-body operators), and it is therefore given roughly by

$$M(s, C) = M_0 - s\delta + C\Delta \qquad (3.9)$$

where $\delta \approx 0.18$ GeV and $\Delta \approx 1.2$ GeV, this estimate for $\Delta$ coming from a comparison between the $\psi$ and $\phi$ masses, giving $(\Delta-\delta)\approx(m(\psi)-m(\phi))/2$. These estimates would suggest a mass of about 2.2 GeV for the lightest baryon with $C\neq 0$. A somewhat higher estimate would be obtained if (3.9) were replaced by a quadratic mass relation, and the corresponding $(\Delta-\delta)$ were estimated from $(m(\psi)^2-m(\phi)^2)/2$; the lightest $C\neq 0$ baryon would then have mass about 3.0 GeV. More detailed structure in the mass spectrum will arise from the properties of the c-(u, d, s) and c-c interactions, their strength and their spin-dependence, their SU(3)-breaking character and their spin-orbit character, as well as on the corresponding properties for the (u, d, s)-(u, d, s) interactions which are already effective in the C=0 baryonic states. Expressions for these effects on the mass spectrum for the lowest supermultiplet of states, with configuration $(1s)_\lambda(1s)_\rho$, have already been provided by Hendry and Lichtenberg.[33]

In these remarks, we have assumed that the c quark moves in the same orbitals as do the (u, d, s) quarks. This would be the case if SU(4) symmetry held for the four quarks q and for the q-q forces, but even if SU(4) symmetry held for the q-q forces, the large mass difference required between the c quark and the (u, d, s) quarks by the comparison of the $\psi$ meson with the $\phi$ and $\rho$ mesons might well cause their orbitals to differ significantly.[34] However, it has been found in various physical situations that there is a general tendency for the space wavefunctions of the low-lying states of few-particle systems to have a greater degree of permutation symmetry than holds for the binding forces themselves. On this basis, it would appear a reasonable first approximation to calculate the baryon mass matrix taking the low-lying orbitals $(n_{12}\ell_{12})_\rho(n_3\ell_3)_\lambda$ to be the same for all four quarks, i.e. to use base wavefunctions which are the same as SU(4) symmetry would require. Further, if the spin-dependence of the c-(u, d, s) central interaction is not too strong, the base supermultiplets may be taken to be those for an SU(8) symmetry,[33] acting in the space of SU(4) x $SU(2)_\sigma$. The SU(8) representations appropriate to the qqq system are 120-, 168-, and 56-dimensional; these correspond to the permutation symmetries S, M and A, respectively, and their SU(4) x $SU(2)_\sigma$ content is as follows:[35]

$$\underline{\underline{120}} = {}^4\underset{\sim}{20}_S + {}^2\underset{\sim}{20}_M, \qquad \underline{\underline{56}} = {}^4\bar{\underset{\sim}{4}}_A + {}^2\underset{\sim}{20}_M,$$

$$\underline{\underline{168}} = {}^4\underset{\sim}{20}_M + {}^2\underset{\sim}{20}_S + {}^2\underset{\sim}{20}_M + {}^2\bar{\underset{\sim}{4}}_A. \tag{3.10}$$

From expressions (3.8), we see that the $\underline{\underline{120}}$ representation has the states $^4(10)$ and $^2(8)$ for C=0, while the $\underline{\underline{168}}$ representation has the states $^4(8)$, $^2(10)$, $^2(8)$ and $^2(1)$ for C=0. These states are the $^{2S+1}(\alpha)$ content of the $\underline{56}$ and $\underline{70}$ representations of SU(6), respectively; the $\underline{\underline{120}}$ and $\underline{\underline{168}}$ representations are simply the direct generalization of the $\underline{56}$ and $\underline{70}$ representations required to include the c quark. In the same way, the $\underline{\underline{56}}$ representation generalizes the $\underline{20}$ representation, for which no baryon is yet established.

These qqq configurations with C≠0 can also undergo internal excitations. Again, with the harmonic shell model, the first excited configuration will have LP=1-, which requires M symmetry. This must be taken together with the SU(8) representation which has M symmetry, namely the $\underline{\underline{168}}$ representation, and so we have ($\underline{\underline{168}}$, 1-) for the first excited supermultiplet, the ground supermultiplet being denoted by $(\underline{\underline{120}}, 0+)_0$. Then there are at least the Regge recurrences and radial excitations of these configurations, and then the other types of excitation discussed already for the C=0 baryonic states. In fact, in the limit of SU(8) symmetry with an harmonic quark shell model, the pattern of supermultiplets expected is exactly that given on Fig.4(b), where the entries dot, square and cross there are now to be identified with the SU(8) representations $\underline{\underline{120}}$, $\underline{\underline{168}}$ and $\underline{\underline{56}}$ respectively. When the quark mass differences are turned on, this pattern of supermultiplets will become greatly distorted on a mass plot, because the coefficient $\Delta$ in Eq.(3.9) is so much larger than the coefficient $\delta$. However, since the additive quantum numbers s and C are both conserved by the strong interactions, the patterns of SU(3) multiplets for definite values (C, s) will still remain recognizable despite the fact that each supermultiplet has now become stretched out over a wide range of masses by this simple SU(4)-breaking mechanism. Other

SU(4)-breaking mechanisms (e.g. through the q-q potentials) can cause mixing between the SU(3)-multiplets for the same (J, P, s, C) and between different SU(8) x O(3) supermultiplets (e.g. between the $(\underline{\underline{120}}, 2+)_2$ and $(\underline{\underline{168}}, 2+)_2$ supermultiplets), in complete analogy with the situation known already for SU(6) x O(3) supermultiplets. The main difference is that the mass differences breaking SU(8) symmetry are an order of magnitude greater than those breaking SU(6) symmetry. However, it is still possible that the major SU(8)-breaking interactions are these mass differences (the "one-body operators") and that the q-q forces at short distances are still well-approximated by SU(8) symmetry, in which case the SU(8)-mixing effects would be very much less than the comparison of $\Delta$ with $\delta$ might suggest.

Consider first the lowest baryon states with C=+1. Whether the $^4(6)_1$ states, the $^2(6)_1$ states or the $^2(\bar{3})_1$ states of the $(\underline{\underline{120}}, 0+)_0$ supermultiplet lie lowest will depend on the details of the SU(8)-breaking, SU(6)-breaking, SU(4)-breaking and SU(3)-breaking character of the q-q interactions. However, the limit of stability for baryons $B_1(s)$ will be given by the mass of $(B_0(s)+D)$, where $B_C(s)$ denotes the lightest baryon for given C and s, and D denotes the lightest C=+1, s=0 meson. With the mass estimates given above, the baryon $B_1(s=0)$ will be stable with respect to $(B_0(0)+D_V)$, whether the linear or quadratic mass formula is used, where $D_V$ denotes the vector D-meson. The pseudoscalar particle $D_p$ is expected to be somewhat lighter than $D_V$. If the quadratic mass formula were used for the baryons in place of (3.9), it is conceivable that the lightest C=+1 baryons $B_1(0)$ could be unstable with respect to strong interactions, but it is really not possible to predict whether or not this should be the case, on the basis of such crude arguments. With the linear mass formula (3.9), on the other hand, the baryons $B_1^*(s)$ would be stable relative to $(B_0(s)+D)$ up to quite high excitation energies above $B_1(s)$. However, $B_1^*(s)$ will be unstable relative to $(B_1(s)+\pi)$ for excitation energy above $m_\pi = 0.14$ GeV. Even for $B_1^*(s)$ below this threshold, rapid decay to $(B_1(s)+\gamma)$ will occur as a process of electromagnetic strength, whenever there is a state $B_1(s)$ with the same charge as the excited state $B_1^*(s)$ under consideration. Similar remarks may be made for baryonic states with higher values for C. Thus, in general, only the lightest baryon $B_C(s)$ for each value of (C, s, Q) can be expected to be semi-stable, decaying through weak interactions processes.

De Rujula et al.[36] have recently made detailed calculations of the masses for members of the $(\underline{\underline{120}}, 0+)_0$ supermultiplet on the basis of the q-q interactions given by a gauge theory for the SU(3)' colour symmetry and incorporating the notions of "asymptotic freedom" and "quark confinement". With the assumption of SU(3) symmetry, there are no free parameters in their q-q interactions, after these have been fitted to the C=0 baryonic masses, other than the c-quark "mass" which is estimated from the $\psi/J$ particle. For C=+1, their calculations place the ${}^2(\bar{3})$ multiplet lowest, the mass values being 2200 MeV and 2420 MeV, for the multiplets with (s, I) = (0, 0) and $(-1, \frac{1}{2})$, respectively. Next comes the ${}^2(6)$ multiplet, the mass values being 2360 MeV, 2510 MeV and 2680 MeV for (s,I) = (0, 1), $(-1, \frac{1}{2})$ and (-2, 0), respectively. The J=3/2 multiplet ${}^4(6)$ lies highest, with mass values 2420 MeV, 2560 MeV and 2720 MeV, for (s, I) = (0, 1), $(-1, \frac{1}{2})$ and (-2, 0), respectively. The semi-stable states for C=+1 all have J=1/2. They are the isosinglet $B_1(0)^+$ and the isodoublet $(B_1(-1)^0, B_1(-1)^+)$ from the ${}^2(\bar{3})_1$ multiplet, and the two isotriplet states $B_1(0)^{++}$ and $B_1(0)^0$ and the isosinglet $B_1(-2)^0$ from the ${}^2(6)_1$ multiplet. These states will all decay through the weak interactions, following the selection rules given in Eqs.(3.13-16) below. The C=+2 and C=+3 states predicted by De Rujula et al. for the $(\underline{\underline{120}}, 0+)_0$ supermultiplet are indicated on Fig.5.

Much discussion has appeared in the literature,[31] concerning the weak decay processes to be expected for the D and S mesons, both vector and pseudoscalar, and for the C≠0 baryons. These weak interactions are believed to result from a coupling of the general form $(GJ_\mu^+ J_\mu)$ between weak interaction currents. These currents $J_\mu$ induce quark transitions with the following properties:

$$(u, d) \to (u, d), \quad \text{with } \Delta C=0,\ \Delta s=0,\ \Delta Q=0, -1, \tag{3.11a}$$

$$(u, d) \to \lambda, \quad \text{with } \Delta C=0,\ \Delta s=-1,\ \Delta Q=-1, \tag{3.11b}$$

$$c \to (u, d), \quad \text{with } \Delta C=-1,\ \Delta s=0,\ \Delta Q=0, -1, \tag{3.11c}$$

$$c \to \lambda, \quad \text{with } \Delta C=-1,\ \Delta s=-1,\ \Delta Q=-1, \tag{3.11d}$$

as well as the corresponding lepton transitions

$$\nu_\ell \to \ell^-, \quad \text{with } \Delta Q=-1, \tag{3.12a}$$

$$\nu_\ell \to \nu_\ell, \text{ and } \ell \to \ell \quad \text{with } \Delta Q=0, \tag{3.12b}$$

and the conjugate currents $J_\mu^+$ induce the charge-conjugate transitions in the reverse direction. In particular models, the various hadron currents may appear with related amplitudes. For example, in the theory advocated by Glashow et al.,[28] the $\Delta Q=-1$ transitions (3.11) appear with the relative amplitudes $\cos\theta_C$: $\sin\theta_C$: $(-\sin\theta_C)$: $\cos\theta_C$, in the order listed, where $\theta_C$ denotes the Cabibbo angle. The $\Delta Q=-1$ leptonic current is believed to be independent of lepton type.

The resulting hadron decay processes may then be semi-leptonic, involving a hadron current (3.11) and a lepton current (3.12) and leading to ($\ell^+ + \nu_\ell$+hadrons), or non-leptonic, involving only the hadron currents (3.11) and their conjugates and leading to hadrons only. For the D and S mesons, the consensus conclusion appears to be that the non-leptonic decay processes will dominate, giving lifetimes of order $10^{-13}$ sec. However, after examination of many possible "all-charged" final states for D meson decay, no evidence has been found for the production and decay of D mesons in $e^+e^-$ collisions at c.m. energy 4.8 GeV,[37] which is above the $D_V\bar{D}_V$ threshold for the $D_V$ mass estimates given above. One difficulty for experiment is that there is a large number of possible non-leptonic decay modes, and further, that only those decay modes where all the final particles are charged (or otherwise visible, such as $K_S^0$ mesons undergoing the charged decay mode $\pi^+\pi^-$) can contribute to the search for these particles. For leptonic decay modes, the situation is even more difficult since the dominant modes are those which arise through the charged lepton current and which therefore include a neutrino. For the semi-stable C=+1 baryons $B_1$(s), again the general expectation is that non-leptonic decay modes will dominate. The currents (3.11) allow the following transitions:

(i) $\Delta C=-1$, $\Delta s=0$, e.g. as follows,

$$B_1(0) \to \pi N, \text{ or } \pi\Delta, \text{ or } \pi\pi N, \text{ etc.} \tag{3.13a}$$

$$B_1(-1) \to \bar{K}N, \text{ or } \pi\Lambda, \text{ or } \pi\Lambda(1520), \text{ or } \bar{K}\pi N, \text{ etc.} \tag{3.13b}$$

(ii) $\Delta C=-1$, $\Delta s=+1$, e.g. as follows,

$$B_1(0) \to KN, \text{ or } K\Delta, \text{ etc.} \tag{3.14a}$$

$$B_1(-1) \to \pi N, \text{ or } \pi\pi N, \text{ or } K\Lambda, \text{ etc.} \tag{3.14b}$$

(iii) $\Delta C=-1$, $\Delta s=-1$, e.g. as follows,

$$B_1(0) \to \bar{K}N, \text{ or } \pi\Lambda, \text{ or } \pi\Sigma(1385), \text{ etc.} \tag{3.15a}$$

$$B_1(-1) \to \pi\Xi, \text{ or } \bar{K}\Lambda, \text{ or } \bar{K}\bar{K}N, \text{ etc.} \tag{3.15b}$$

(iv) $\Delta C=0$, $\Delta s=+1$, which we know already from the decay processes of the C=0 baryons and which will give rise to corresponding processes for the C=+1 baryons, such as

$$B_1(-1) \to \pi B_1(0). \tag{3.16}$$

The relative rates for all these decays will naturally depend on the details of the weak interactions, concerning which present views are in a rather fluid state.

The lightest $B_1(s)$ baryons will therefore be seen as narrow resonances, with decay widths of order $10^{-2}$ eV corresponding to typical lifetimes of order $10^{-13}$sec, in many final hadronic states. None have been established to date and they pose a problem of a different order of magnitude of difficulty in comparison with current investigations of the $\psi$ particles. From the above remarks, we see that they can, in principle, be formed directly in meson-baryon collisions. For example, from Eqs. (3.13-16), the baryons $B_1(0)$ can be formed in $\pi N$, $KN$ and $\bar{K}N$ collisions, and $B_1(-1)$ can be formed in $\pi N$ and $\bar{K}N$ collisions. However, the integrated cross sections for these formation processes are negligibly small relative to the cross section for the non-resonant scattering integrated over the energy resolution $\Delta E$; for $J=1/2$, the former is $4\pi\lambda^2\Gamma_{\pi N}$, to be compared with $4\pi\lambda^2\nu(\Delta E)$ where $\nu \lesssim 1$ measures the strength of the non-resonant scattering relative to the geometric limit and is typically of order $10^{-2}$, while $\Delta E$ is of order 1 MeV compared with $\Gamma_{\pi N} \approx 10^{-7}$ MeV. The situation would be rather comparable with attempting to establish the existence of the $\Lambda$ particle by studying s-wave $\pi N$ scattering in the neighbourhood of incident pion laboratory momentum 119.3 MeV/c.[38] The $e^+e^- \to \psi$ process has the great advantage that the background reactions are only electromagnetic in origin and with cross section of order $(1/137)^2 4\pi\lambda^2$, while the convenient entrance channel $e^+e^-$ has a substantial branching ratio, $\Gamma(\psi\to e^+e^-)/\Gamma(\psi\to \text{all}) \approx 0.07$, with $\Gamma(\psi\to \text{all}) \approx 80$ keV being about one-tenth of the energy resolution for the electron and positron beams circulating in the storage ring. In this case, we have to compare $\Gamma(\psi\to e^+e^-) \approx 5$ keV with $(1/137)^2\Delta E \sim 5 \times 10^{-2}$ keV, a very favourable ratio.

Comparison with the case of strangeness is instructive. Here, the most productive studies have been for formation experiments $\bar{K}N \to \Lambda^*$ and $\Sigma^*$, where the resonance state of interest is formed with quite a low velocity in the laboratory frame, as has also been the case in the study of $N^*$ and $\Delta^*$ resonance states. For the quantum number C, on the other hand, we lack any $C\neq 0$ meson with lifetime sufficiently long to permit the construction of a $C\neq 0$ meson beam for cross section measurements and resonance formation. A lifetime $\tau \sim 10^{-13}$ sec for the D meson allows a mean path length of only $30(E_D/m_D)$ microns and the probability for a nuclear collision in this distance through normal matter is rather small. In assessing the situation, it is a chastening thought to compare our state of knowledge concerning $N^*$, $\Delta^*$, $\Lambda^*$ and $\Sigma^*$ resonances, where formation experiments are possible, with our meagre knowledge of $\Xi^*$ and $\Omega^*$ resonances, where our information can come only from production experiments.

However, we turn now to consider production processes and mention briefly three classes of experiment.

(i) Associated Production. The direct analogues of the $\pi^- P \to \Lambda K^0$ reaction through which we gained a large part of our early knowlege about strange particles, are the processes

$$\pi^- + P \to \bar{D}^0 + B_1^0(0), \tag{3.17a}$$

$$\to \bar{D}^{--} + B_1^+(0), \tag{3.17b}$$

$$\pi^+ + P \to \bar{D}^0 + B_1^{++}(0), \tag{3.17c}$$

for $\bar{D}$ and $B_1(0)$ production, where we note that there are three states $B_1^+(0)$ belonging to the ground supermultiplet $(\underline{\underline{120}}, 0+)_0$, two with J=1/2 belonging to the $^2(6)_1$ and $^2(\bar{3})_1$ unitary multiplets and one with J=3/2 belonging to the $^4(6)_1$ unitary multiplet, and the processes

$$K^- + P \to \bar{D}^0 + B_1^0(-1), \tag{3.18}$$

for $B_1(-1)$ production, where we note that there are three states $B_1^0(-1)$ belonging to $(\underline{\underline{120}}, 0+)_0$, two with J=1/2 belonging to the $^2(6)_1$ and $^2(\bar{3})_1$ unitary multiplets and one with J=3/2 belonging to the $^4(6)_1$ unitary multiplet. Of course, these $B_1(s)$ production reactions (3.17) and (3.18) may proceed with associated production of vector or pseudoscalar $\bar{D}$ mesons, or of $\bar{D}$ mesons with higher spin values. Their thresholds will lie high, at laboratory $\pi$ or $\bar{K}$ momentum of 10 GeV/c or more. A reasonable expectation would be that the cross section rises rapidly near threshold to some fraction (perhaps as high as 1% - probably much less since there are very many channels open at this energy) of $\pi\lambda^2 \approx 0.3$ mb, the geometric limit on the s-wave reaction cross section at this energy, and then falls away gradually as the energy rises further - thus giving a maximum cross section of no more than about 1 μb, as a reasonably optimistic estimate.

(ii) Pair Production. The most favourable reaction for experimentation is with an electron-positron storage ring, namely:

$$e^+ + e^- \to D^+ + \bar{D}^{--}, \qquad (3.19a)$$

$$\to S^+ + \bar{S}^{-}. \qquad (3.19b)$$

The energies at present available are more than adequate for these processes. Although the cross sections are small, being electromagnetic, it is reasonable to expect these cross sections to be a non-negligible fraction of the total cross section, not far above threshold. The major uncertainty is whether the $\gamma DD$ form factor may reduce the cross section far below simple expectation. However, at SPEAR, $p\bar{p}$ and $\Lambda\bar{\Lambda}$ pair production (with associated pions) has already been observed,[24] with a branching ratio of order 1%, at the $\psi/J(3.1)$ resonance. It is interesting to reflect that $\Omega\bar{\Omega}$ production may similarly be expected to have a comparable branching ratio, perhaps of order 0.1%, at the $\psi(3.7)$ resonance; in this event, the SPEAR storage ring would ultimately provide $\Omega$ and $\bar{\Omega}$ particles for experimental study in a most favourable situation, namely (i) producing them with an event/background ratio which appears rather high, when compared with the circumstances of $\Omega$ production in high energy $K^-p$ interactions, and (ii) producing them with a laboratory momentum which is not highly relativistic. It seems quite a hopeful possibility that one may be able conveniently to carry out work in C=+1 baryon spectroscopy (and perhaps for C=+2, and higher), as well as baryon spectroscopy for C=0, with electron-positron storage rings, especially if further narrow resonances are found in work at still higher energies.

Pair production may also be studied in other interactions, such as

$$P + P \to P + P + B_1(s) + \bar{B}_1(-s). \qquad (3.20)$$

With proton beam on stationary target, the threshold is high, at incident momentum ≈20 GeV/c, and the C=+1 baryon and C=−1 antibaryon are produced with highly relativistic momentum, and with a low branching ratio, as we know already for $P\bar{P}$ production. Of course, with a proton storage ring for proton energy 20 GeV, far more than enough energy is available for $B_1\bar{B}_1$ pair production and the difficulty is rather than the relevant events are swamped by the high yield of pions and other light particles in these PP collisions.

(iii) Neutrino-induced Reactions. Since the weak interaction violates almost all of the non-geometric selection rules we know in hadronic physics (leaving only baryon number and charge as rigorously conserved quantum numbers), neutrino reactions are especially favourable for the production of particles with new quantum numbers. It will be sufficient to give one illustration, as follows. The reaction

$$\nu + P \rightarrow \mu^- + B_1^{++}(0), \tag{3.21}$$

is already energetically possible for rather low neutrino energy, but a much higher neutrino energy is needed before the cross section becomes reasonably large for experiment. The $B_1^{++}(0)$ produced may have either J=3/2, as a member of the $^4(6)_1$ unitary multiplet, or J=1/2, as a member of the $^2(6)_1$ unitary multiplet, of the $(\underline{\underline{120}}, 0+)_0$ supermultiplet.

It is remarkable that one neutrino-induced event has recently been reported[39] which could be an example of this production reaction (3.21). The event is fitted as

$$\nu + P \rightarrow \mu^- + \Lambda + \pi^+ + \pi^+ + \pi^+ + \pi^-, \tag{3.22}$$

with ($\Lambda\pi\pi\pi\pi$) mass 2426 MeV. The interpretation suggested by the authors on the basis of the calculations by De Rujula et al.[36] for the $(\underline{\underline{120}}, 0+)_0$ supermultiplet, which we described briefly above, is that this event is an example of the $B_1(0)^{++}$ production process (3.21), for the $B_1(0)^{++}$ state in its $^4(6)$ sub-multiplet, followed by the strong hadronic decay process

$$B_1(0, {}^4(6))^{++} \rightarrow \pi^+ + B_1(0, {}^2(\bar{3}))^+. \tag{3.23}$$

This last state has mass 2200 MeV, according to De Rujula et al., and is one of the semi-stable C=+1 states mentioned above; its decay will then have proceeded through the non-leptonic weak interaction with $\Delta C=-1$, $\Delta s=-1$ and $\Delta Q=0$, resulting from the combination of the weak currents (3.11a) and (3.11d), which gives

$$B_1(0, {}^2(\bar{3}))^+ \rightarrow \pi^+ + (\Lambda^* \text{ or } \Sigma^{*0}). \tag{3.24}$$

Although a plausible possibility, this interpretation cannot be demonstrated, on the basis of one event. Nevertheless, this event does illustrate well the kind of event which would result from neutrino excitation of a C=+1 baryonic state, followed by its decay through strong, electromagnetic or weak decay processes. In the course of time, we may expect a gradual collecting of statistics for the very many reaction processes which can result from the excitation of C=+1 baryons by neutrinos and their subsequent decay and which may be used to demonstrate their existence, their mass values and their unitary multiplet patterns. The event rate for specific configurations will be exceedingly low, however, and it is likely to be a long time before any spin-parity values for C=+1 baryons will be determined from such data.

## References


1. See, for example, J. Jost, "Das Pauli-Prinzip und die Lorentz-Gruppe" in "Theoretical Physics in the Twentieth Century" (eds. M. Fierz and V. F. Weisskopf, Interscience Publ. Inc., New York, 1960), p.107.

2. M. Gell-Mann, in Elementary Particle Physics (ed. P. Urban, Springer-Verlag, Vienna, 1972), p.733.

3. H. J. Lipkin, Phys. Letters 45B (1973) 267.

4. R. H. Dalitz, in Proc. Second Hawaii Topical Conf. in Particle Physics (1967), (eds. S. Pakvasa and S. F. Tuan, Univ. Hawaii Press, Honolulu, Hawaii, 1968). See p.398.

5. Any finite number of individual states may be absent from a given Regge trajectory. The absence of a state simply means a zero residue at the Regge pole which would correspond to that state.

6. S. Frautschi, Regge Poles and S-Matrix Theory (W. A. Benjamin Inc., New York, 1963).

7. The suffix n used henceforth in the SU(6)-supermultiplet notation $(\underline{N}, LP)_n$ gives the degree of excitation n=(L+2r), as in Eq.(1.6).

8. O. W. Greenberg and M. Resnikoff, Phys. Rev. 163 (1967) 1844.

9. G. Karl and E. Obryk, Nucl. Phys. B8 (1968) 609.

10. P. J. Litchfield, in Proc.XVII Int. Conf. on High Energy Physics (ed. J. R. Smith, Science Research Council, Rutherford Laboratory, Chilton, 1974), p.II-94.

11. R. J. Cashmore, A. J. G. Hey and P. J. Litchfield, Nucl. Phys. B98 (1975) 237.

12. R. Horgan, Nucl. Phys. B71 (1974) 514.

13. Particle Data Group, Phys. Letters 50B (1974) 1.

14. For further references on this topic, see M. Moshinsky, The Harmonic Oscillator in Modern Physics: From Atoms to Quarks (Gordan and Breach, New York, 1969).

15. M. Moshinsky[14] has directed attention to the possibility of a unique characterization for each of these states in terms of the SU(3(n-1)) group of the Hamiltonian describing n particles interacting through particle-particle harmonic potentials.

16. For the SU(3) group appropriate to the motions of a three-particle system, a complete set of commuting operators, sufficient to characterize each state uniquely, has been given by B. R. Judd, W. Miller, J. Patera and P. Winternitz in J. Math. Phys. 15 (1974) 1787.

17. A simple and direct approach to the unique characterization of all the states of a three-particle system moving under particle-particle harmonic potentials has been worked out by R. R. Horgan at Oxford. See R. R. Horgan, "The Construction and Classification of Wavefunctions for the Harmonic Oscillator Model of Three Quarks", Oxford preprint (1976).

18. D. B. Lichtenberg, Phys. Rev. 178 (1968) 2197.

19. S. Ono, Progr. Theor. Phys. 48 (1972) 964.

20. R. H. Capps, Phys. Rev. Letters 33 (1974) 1637; Phys. Rev. D12 (1975) 3606.

21. T. Eguchi, "Baryons, Diquarks and Strings", Enrico Fermi Institute (University of Chicago) preprint EFI 75-50 (September 1975).

22. A. N. Mitra, Phys. Letters 51B (1974) 149.

23. A. N. Mitra, Phys. Rev. D11 (1975) 3270.

24. Particle Data Group, Rev. Modern Phys. 47 (1975) 535.

25. R. F. Schwitters, Proc.1975 Intl.Symposium on Lepton and Photon Interactions at High Energies (ed.W. T. Kirk, SLAC, Stanford Univ., California), p.5.

26. See the review papers by G. J. Feldman, B. H. Wiik and J. Heintze in Proc.1975 Intl. Symposium on Lepton and Photon Interactions at High Energies (ed. W. T. Kirk, SLAC, Stanford Univ., California) on pages 39, 69 and 97, respectively.

27. J. D. Bjorken and S. L. Glashow, Phys. Letters 11 (1964) 255.

28. S. L. Glashow, J. Iliopoulos and L. Maiani, Phys. Rev. D2 (1970) 1285.

29. G. Zweig, in "Symmetries in Elementary Particle Physics" (Academic Press, New York, 1965), p.192.

30. The charge assignment which is physically correct for the quark c will become settled directly when the charge states are determined for the hadrons having C≠0.

31. M. Gaillard, B. W. Lee and J. L. Rosner, Revs. Modern Phys. 47 (1975) 277.

32. From the study of μe events in $e^+e^-$ annihilation, Perl et al. (see M. Perl, "Properties of eμ Events Produced in $e^+e^-$ Annihilation", SLAC preprint SLAC-PUB-1664, November 1975), have made a powerful case for the existence of heavy leptons U of mass between 1.8 and 2.0 GeV. If this proves correct, then an increase in R by +1 will occur over this energy range due to the onset of $e^+e^- \to U^+U^-$ pair creation, but this is less than 50% of the increase observed in R.

33. A. W. Hendry and D. B. Lichtenberg, Phys. Rev. D12 (1975) 2756.

34. This remark assumes "light quarks", as is commonly accepted at present. If the truth corresponds to "heavy quarks", these remarks no longer follow, since these mass differences would then be small relative to the mean quark mass.

35. J. W. Moffat, Phys. Rev. 140B (1965) 1681; S. Iwao, Ann. Phys. (N.Y.) 35 (1965) 1.

36. A. De Rujula, H. Georgi and S. L. Glashow, Phys. Rev. D12 (1975) 147.

37. A. Boyarksi *et al.*, Phys. Rev. Letters 35 (1975) 195.

38. This is the incident pion momentum for which the reaction $\pi^- + p \to \Lambda$ occurs for a stationary proton target.

39. E. Cazzoli, A. Cnops, P. Conolly, R. Loutit, M. Murtagh, R. Palmer, N. Samios, T. Tso and H. Williams, Phys. Rev. Letters 34 (1975) 1125.

# NULL PLANE FIELD THEORY AND COMPOSITE MODELS

J. S. Bell
CERN -- Geneva

and

H. Ruegg *)
University of Geneva

ABSTRACT

A sketch is given of an approach to null plane field theory which (it is hoped) illuminates the relation between the relativistic parton model, the non-relativistic quark model, and various SU(6) and $SU(6)_W$ broken symmetry schemes, including those of Melosh.

## 1. INTRODUCTION

This is a sketch of an approach to null plane field theory [1] designed to meet the following requirements :

1) to exhibit, quite explicitly, the null plane approach as a method in ordinary quantum field theory, as distinct from a fresh departure from the classical Lagrangian;

*) Partially supported by the Swiss National Fund.

2) to make it easy to pass to a non-relativistic approximation, and so hold closely together in one formalism the relativistic parton model and the non-relativistic quark model;

3) to extend to interacting field theory [2] considerations on the Melosh transformations previously made for free field theory [3], for a potential model [4], and for a quasi-potential model [5].

None of these requirements has been definitely and convincingly met. This is a preliminary account, to an informal meeting, of undigested ideas.

## 2. COMPOSITE SYSTEMS ON THE NULL PLANE

There is a naïve way of regarding complex systems in field theory as made up of non-interacting particles. Imagine the interaction switched off at some instant. The state then indeed becomes one of free particles, and this free particle state can be used to characterize the complex system. In this sense the physical vacuum at zero time is represented by

$$T\, e^{-i\int d^4x\, \Theta(-t)\, \mathcal{H}_I(x)}\, |\text{bare vacuum}\rangle \tag{1}$$

where $T$ denotes time-ordering, $\Theta$ is the step function, $\mathcal{H}_I$ is the interaction Hamiltonian in the interaction representation, and an adiabatic switch-on is implicit. To deal with states other than vacuum suitable source terms can be added to $\mathcal{H}_I$.

In this picture even the vacuum is complicated. The instantanenous switch-off does not conserve energy, so that even the vacuum gives rise to a sea of bare particles and antiparticles.

However, why switch off everywhere at the same time ? There is nothing sacred about simultaneity. Let us switch off instead on a null plane, taking instead of (1)

$$T\, e^{-i\int d^4x\, \Theta(-t-z)\, \mathcal{H}_I(x)}\, |\text{bare vacuum}\rangle \tag{2}$$

Again energy is not conserved, and now also the third component of momentum is not conserved. But because the switch-off is invariant under simultaneous displacement in time and along the $z$ direction in space, the <u>sum</u> of energy and momentum (in the $z$ direction) is conserved. So a state containing particles with momenta $p$ can arise from the vacuum only if

$$\sum (p_0 + p_3) = 0$$

where

$$p_0 = \sqrt{m^2 + \vec{p}^2} > |p_3|$$

(excluding the case of zero mass). Since all the terms in the sum are positive only the zero particle state can arise from the vacuum.

In this picture real and bare vacua are identical. This is very convenient for composite models, since the constituents of the composite do not have to be disentangled from those of the vacuum.

The null plane switch off is not rotation invariant (except for rotation about the 3-axis). The description of rotation is correspondingly complicated, becoming interaction dependent. On the other hand, the null plane switch-off is invariant for Lorentz boosts in the 3-direction. The description of this part of Lorentz invariance is correspondingly simplified.

## 3. NULL PLANE DYNAMICS

We can make (2) look more like (1) by introducing

$$\hat{\mathcal{L}}_I(t, x, y, z) = \mathcal{L}_I(t-z, x, y, z) \tag{3}$$

Then (2) becomes

$$T\, e^{-i\int d^4x\, \theta(-t)\, \hat{\mathcal{L}}_I(x)}\, |\text{bare vacuum}\rangle \tag{4}$$

at least if

$$[\mathcal{L}_I(x), \mathcal{L}_I(y)] = 0 \qquad \text{for } |x_0 - y_0| \lesssim |\vec{x} - \vec{y}| \tag{5}$$

These commutators arise because of the reordering in going from (2) to (4); ordering operators $\hat{\mathcal{L}}_I(t,x,y,z)$ with respect to $t$ is not the same as ordering operators $\mathcal{L}(t,x,y,z)$. We will tentatively suppose (4) to be equivalent to (2), at least in some suitably regulated version of the theory.

Comparing (4) and (2), we see that null plane dynamics is the same as equal time dynamics but with a different Hamiltonian. All the usual quantum mechanical formalism is therefore available. When $\mathcal{H}_I$ is a combination of field variables $\psi$, $\hat{\mathcal{H}}_I$ is the same combination of new fields

$$\hat{\psi}(t,x,y,z) = \psi(t-z,x,y,z) \tag{6}$$

Since there is only a change of time argument here, the $\hat{\psi}$ satisfy the same equations of motion

$$i\frac{\partial\hat{\psi}}{\partial t} = [\hat{\psi}, H_0] \tag{7}$$

where $H_0$ is the same, original, unmodified, constant, free Hamiltonian. From the interaction picture we can pass to a Schroedinger picture in the usual way, taking $t=0$ in the free fields $\hat{\psi}$. The complete Hamiltonian is

$$\hat{H} = H_0 + \int d^4x\, \delta(t)\, \hat{\mathcal{H}}_I(x) \tag{8}$$

From this Schroedinger picture we can pass in the usual way to a Heisenberg picture. As usual the bound states of composite systems will appear as eigenstates of the total Hamiltonian (8).

Note, however, an important difference of detail between null plane and equal time dynamics. The Hamiltonian (8) does not conserve z momentum, but rather the combination of z momentum and energy :

$$\Delta \sum (p_0 + p_3) = 0 \tag{9}$$

It is usual to write

$$\eta = p_0 + p_3 \tag{10}$$

and to denote the other components of momentum by $\vec{p}_\perp$. Then the Hamiltonian connects states such that

$$\Delta \sum \vec{p}_\perp = \Delta \sum \eta = 0 \tag{11}$$

## 4. NON-RELATIVISTIC QUARK MODEL AND RELATIVISTIC PARTON MODEL

In view of the success of non-relativistic potential models it may be useful to write the null plane Hamiltonian in the form

$$\hat{H} = H_0 + V + \left( \int d^4x \, \delta(t) \, \hat{\mathcal{H}}_t(x) \quad - V \right) \tag{12}$$

where V is a potential, i.e., it scatters particles (or antiparticles) on one another but does not produce or absorb them. In null plane dynamics it is natural to choose V such that

$$\Delta \sum h = 0 \tag{13}$$

rather than

$$\Delta \sum p_3 = 0 \tag{14}$$

But when all particles move slowly, i.e.,

$$h = p_3 + m + o(c^{-2})$$

and when particles are not absorbed or produced, (13) implies (14) apart from relativistic corrections. Neglecting in $H_0$ relativistic corrections to particle energy

$$p_0 \approx m + \vec{p}^{\,2}/2m$$

the first two terms of (12) give a non-relativistic potential model.

When the remaining terms of (12) are taken into account, either by formal perturbation theory or more realistically, eigenstates $\Psi$ of the full Hamiltonian are mapped onto eigenstates $\Psi'$ of the model :

$$\Psi = F \Psi' \tag{15}$$

The linear operator F must create the gluons and particle-antiparticle parts contained in the real state $\Psi$ and not in the model state $\Psi'$ – in the relativistic parton model but not in the non-relativistic quark model.

The notion of model state and of the associated transformation to the real state have long been familiar in, for example, nuclear physics, as has been recalled in the quark model context by G. Morpurgo [6].

Matrix elements of an operator, for example a current $J$, can of course be calculated from either state vector :

$$(\Psi, J\Psi) = (\Psi', F^+ J F \Psi') \tag{16}$$

There is a choice between complicated states $\Psi$ and simple operators $J$, on the one hand, or simple states $\Psi'$ and complicated operators $F^+JF$ on the other. For this reason we have suggested [4] that the two pictures $\Psi$ and $\Psi'$ of the same state be called "current" and "constituent" pictures respectively - the first involving relatively simple forms for currents and the latter involving relatively few constituents. The model operator $F$, which transforms between what we call constituent and current pictures, is something quite other than what we call the "Melosh transformation" - which is a way of exhibiting the approximate spin independence of interactions. This distinction has been insisted on also by G. Morpurgo [6].

## 5. SPIN INDEPENDENCE

Interaction matrix elements tend to become spin independent when all the particles involved are slow. That is to say that we have invariance under rotation of the spin states, leaving the momenta fixed. This follows from ordinary rotation invariance; we can omit the rotation of momenta when these are negligibly small. However, it may not be realistic, in some case of interest, to regard <u>all</u> particles as slow. There may be a light meson (or even photon) involved, which has to be considered as relativistic. It may be enough, however, that <u>other</u> particles are slow, as we will see in some examples.

We work here in interaction representation, so that we can exploit knowledge of free field operators (Appendix) and pass readily to null plane dynamics.

Note first the spin independence of the free Dirac Hamiltonian

$$\int d^4x\ \delta(t)\ \psi^+ (\beta m + \vec{\alpha}\cdot\vec{p})\ \psi$$
$$= \int d^4x\ \delta(t)\ \phi^+ \beta \sqrt{m^2+\vec{p}^{\,2}}\ \phi$$

where $\phi$ is the Foldy-Wouthuysen field operator. This is invariant under

$$\delta\phi = -i\Lambda\phi$$

provided

$$[\beta, \Lambda] = 0$$

i.e.,

$$\Lambda = \tfrac{1}{2}(1 \pm \beta)(1, \vec{\sigma})$$

This symmetry will be denoted by

$$(U(2) \times U(2))_{FW} \qquad (17)$$

Moreover, it will be convenient to separate the generators into two sets, defining two subgroups

$$\left.\begin{array}{ll} U(2)_{FW} & \Lambda = (1, \vec{\sigma}) \\ U(2)_{FWW} & \Lambda = (1, \beta\sigma_x, \beta\sigma_y, \sigma_z) \end{array}\right\} \qquad (18)$$

Given more than one fermion field, say three, of equal mass, these groups generalize in an obvious way to $(U(6)\times U(6))_{FW}$, etc.

Add now an interaction

$$\mathcal{H}_I = g\bar{\psi} A \psi \qquad (19)$$

where A is a neutral scalar gluon. In terms of the Foldy-Wouthuysen field $\phi$,

$$\psi = \left(1 - \frac{i\vec{\gamma}\cdot\vec{p}}{2m} + \cdots\right)\phi \qquad (20)$$

so that to first order in small quantities $(\vec{p}/m)$, $(\vec{p}\,'/m)$,

$$\begin{aligned} g^{-1}\mathcal{H}_I = {} & \phi^{+}\beta A\phi \\ & + \phi^{+}\frac{i\vec{\gamma}\cdot\vec{p}\,'}{2m}\beta\phi - \phi^{+}\beta\frac{i\vec{\gamma}\cdot\vec{p}}{2m}\phi \end{aligned} \qquad (21)$$

where $\vec{p}\,'$ denotes a final momentum, as distinct from an initial momentum $\vec{p}$, or alternatively a differential operator on $\phi^{+}$ rather than $\phi$. The leading term in (21) is invariant under the full $(U(2)\times U(2))_{FW}$. The subsequent terms are not invariant, and it looks as if symmetry breaking sets in at first order.

But it can be argued that for many purposes this is effectively not so. Suppose that we do not have sources or sinks of gluons other than the interaction Hamiltonian (19). Then in the course of a perturbation calculation such terms have to be paired off with one another. Moreover, when all fermions are slow, leading and subleading parts of (21) cannot pair together. This is because the subleading parts necessarily create or absorb a fermion-antifermion pair (because $i\vec{\gamma}\beta$ anticommutes with $\beta$) and therefore (in null plane dynamics) a gluon of large $\eta$ ($\sim 2m$). For the leading part the reverse is true. So the symmetry breaking terms contribute in twos, and the symmetry breaking is effectively of second rather than first order.

Consider now an interaction

$$\mathcal{H}_I = g\, \bar{\Psi}\, i\gamma_\mu A_\mu \Psi \tag{22}$$

with a neutral vector gluon. To first order

$$\begin{aligned} g^{-1}\mathcal{H}_I = & -\phi^+ A_0 \phi \\ & + \phi^+ \vec{\alpha}\cdot\vec{A}\, \phi \\ & - \phi^+ \left( \frac{i\vec{\gamma}\cdot\vec{p}}{2m} A_0 - A_0 \frac{i\vec{\gamma}\cdot\vec{p}}{2m} \right) \phi \\ & + \phi^+ \left( \frac{i\vec{\gamma}\cdot\vec{p}}{2m} \vec{\alpha}\cdot\vec{A} - \vec{\alpha}\cdot\vec{A}\, \frac{i\vec{\gamma}\cdot\vec{p}}{2m} \right) \phi \end{aligned} \tag{23}$$

The discussion of the first and third terms here goes along the lines just given for the scalar gluon, with the same conclusions. The second term here goes along the lines just given for the scalar gluon, with the same conclusions. The second term is more troublesome; it breaks the $(U(2)\times U(2))_{FW}$ symmetry, and looks of similar order to the leading term. The subgroup

$$U(2)_{FW}$$

can be restored by augmenting the infinitesimal operation

$$\delta\phi = -i\vec{\omega}\cdot\vec{\sigma}\,\phi \tag{24}$$

by

$$\delta\vec{A} = \vec{\omega}\times\vec{A} \tag{25}$$

This latter is an invariance of the free Hamiltonian if we adopt the Gupta-Bleuler formalism. More generally, note that the second term in (18), because

$\alpha$ anticommutes with $\beta$, is a fermion-antifermion hard gluon type. Again for many purposes such terms have to be taken in pairs. The emission and absorption matrix elements of the hard gluon each involve a factor $1/\sqrt{\eta}$ and the propagation involves an energy denominator of order $\eta \approx m$ (assuming $\mu \ll m$, where $\mu$ is the gluon mass). So for such contributions there is a suppression factor $\approx m^{-2}$, which has to be compared with $\mu^{-2}$ or $|\vec{p}|^{-2}$ (where $\vec{p}$ is a typical fermion momentum) for leading contributions. In this sense, whose precision clearly leaves much to be desired, $(U(2)\times U(2))_{FW}$ again holds good to first order.

We mention only briefly two other kinds of gluon [4]. With a pseudoscalar gluon and

$$\mathcal{H}_I = g\,\bar{\Psi}\, i\gamma_5 A\, \Psi$$

the leading terms are already fermion-antifermion-hard-gluon-type and correspondingly suppressed. With a pseudovector gluon and

$$\mathcal{H}_I = g\,\bar{\Psi}\, i\gamma_5\gamma_\mu A_\mu\, \Psi$$

the leading terms already break $(U(2)\times U(2))_{FW}$, leaving only $U(2)_{FW}$ in the sense of (24), (25). Moreover, the associated Stueckelberg pseudoscalar has gradient coupling with strength $(g/\mu)$, so that the "small" corrections involve factors $(\vec{p}/\mu)$ as well as $(\vec{p}/m)$.

## 6. NULL PLANE CHARGES

The free Hamiltonian can be expressed in terms of the good components $\hat{\psi}_g$ of the null plane field (see Appendix)

$$H_0 = 2\int d^4x\; \delta(t)\; \hat{\Psi}_g \left(\frac{\eta}{2} + \frac{m^2 + \vec{p}_\perp^{\,2}}{2\eta}\right) \hat{\Psi}_g \qquad (26)$$

It is then clearly invariant under a symmetry

$$U(2)_W \qquad \delta\hat{\Psi}_g = -i\,\Lambda\,\hat{\Psi}_g \qquad \Lambda = (1, \sigma_x, \sigma_y, \sigma_z) \qquad (27)$$

In view of the null-plane canonical anticommutators, this can be considered to be generated by the good null plane charges

$$2\int d^4x\; \delta(t)\; \hat{\Psi}_g^+\, \Lambda\, \hat{\Psi}_g \qquad (28)$$

A term in the interaction Hamiltonian, for example

$$\psi^{+}\psi \; A_{0} \tag{29}$$

can be expressed in terms of $\hat{\psi}_g$ by eliminating the bad components

$$\hat{\psi}_b = \eta^{-1}(m\sigma_3 - \vec{\sigma}_\perp\cdot\vec{p}_\perp)\,\hat{\psi}_g \tag{30}$$

Then (29) becomes

$$\begin{aligned} &\hat{\psi}_g^{+} A_0 \hat{\psi}_g \; + \\ &\hat{\psi}_g^{+}(m\sigma_3 - \vec{\sigma}_\perp\cdot\vec{p}'_\perp)\eta'^{-1} A_0\, \eta (m\sigma_3 - \vec{\sigma}_\perp\cdot\vec{p}_\perp)\,\hat{\psi}_g \end{aligned} \tag{31}$$

When the small quantities $(\vec{p}_\perp/m, \vec{p}'_\perp/m)$ are neglected, the interaction respects the symmetry (27). But this symmetry is broken at first order.

The Foldy-Wouthuysen symmetry of the last section is (for many purposes, we argued) still good at first order. Moreover, there is no reason why it should not be expressed on the null plane. Writing the 4×4 Dirac rotation matrices $\sigma$ in the good representation (A.7) and eliminating reference to bad components by

$$\phi_b = \varepsilon\,\sigma_3\,\phi_g$$

[from (A.9)] where

$$\varepsilon = \text{sign}\, p_0 = \text{sign}\,\eta = \sqrt{\eta^2}/\eta$$

we have the symmetry operations

$$\begin{aligned} &U(2)_{FW} && \delta\hat{\phi}_g = -i\Lambda\hat{\phi}_g && \Lambda = (1, \varepsilon\sigma_x, \varepsilon\sigma_y, \sigma_z) \\ &U(2)_{FWW} && \delta\hat{\phi}_g = -i\Lambda\hat{\phi}_g && \Lambda = (1, \sigma_x, \sigma_y, \sigma_z) \end{aligned} \tag{32}$$

These are generated by the null plane charges

$$2\int d^4x \; \delta(t) \; \phi_g^{+} \Lambda \phi_g \tag{33}$$

These differ from the corresponding constructions, including (28), from the fields $\hat{\psi}$, by

$$\hat{\psi}_g = \frac{m + \sqrt{\eta^2} + \sigma_3\vec{\sigma}_\perp\cdot\vec{p}_\perp}{\sqrt{(m+\sqrt{\eta^2}\,)^2 + \vec{p}_\perp^{\,2}}}\;\hat{\phi}_g \tag{34}$$

- the "second Melosh" transformation of (A.37).

One could of course, using the inverse of (34), rewrite the symmetry operations (32) and (33) in terms of the fields $\hat{\psi}_g$. They would then be rather complicated. On the other hand the operator for a physical quantity like a current

$$\bar{\psi}\, i\gamma_\mu\, \psi$$

is rather complicated in terms of the $\hat{\phi}_g$. So we propose calling $\hat{\psi}_g$ and $\hat{\phi}_g$ "current" and "classification" operators respectively, according to whether they simplify the expression of currents or symmetry operations.

As always in quantum mechanics there is the option, instead of changing operators while holding state vectors fixed, of holding the operators fixed and changing the state vectors. In this latter version [4] the Melosh transformation is between "current" and "classification" pictures. It has nothing to do, in our terminology, with the relation between "current" and "constituent" pictures.

## 7. BOOST INVARIANCE

The good null plane charges (35) are invariant for boosts in the z direction. The Melosh modified charges (33) are not, if the full form (34) of the second Melosh transformation is used, for this transformation itself is not boost invariant. But our symmetries were good only to first order in small quantities. To this order the transformation (34) is equivalent to

$$\hat{\psi}_g = e^{\frac{\sigma_3 \vec{\sigma}_\perp \cdot \vec{p}_\perp}{2m}}\, \hat{\phi}_g = \left(1 + \frac{\sigma_3 \vec{\sigma}_\perp \cdot \vec{p}_\perp}{2m} + \cdots\right) \hat{\phi}_g \qquad (35)$$

- or indeed to any such form with the same first two terms, including the "first Melosh" transformation [7]

$$\hat{\psi}_g = \frac{m + \sqrt{m^2 + \vec{p}_\perp^{\,2}} + \sigma_3 \vec{\sigma}_\perp \cdot \vec{p}_\perp}{\sqrt{\left(m + \sqrt{m^2 + \vec{p}_\perp^{\,2}}\right)^2 + \vec{p}_\perp^{\,2}}}\, \hat{\phi}_g \qquad (36)$$

If (35) instead of (34) is used to define $\hat{\phi}_g$ in terms of $\hat{\psi}_g$, the symmetry operations (32) and the corresponding charges (33) become boost invariant. Moreover, this symmetry, which we might call

$$(U(2)\times U(2))_{FWB}\ ,\quad U(2)_{FWB}\ ,\quad U(2)_{FWWB} \tag{37}$$

is just as good (as far as our arguments go) as

$$(U(2)\times U(2))_{FW}\ ,\quad U(2)_{FW}\ ,\quad U(2)_{FWW}$$

Moreover, when one of the symmetries (37) is good for some composite system, because of slow motion of the constituents

$$(\vec{p}_\perp/m)^2 \ll 1 \qquad |(\eta/m)-1| \ll 1$$

it remains good for that same system arbitrarily boosted in the 3-direction.

## 8. REGULARIZATION

We return now to one of the subtleties that have been trampled over, to trample less heavily. Consider again the reordering of operators in passing from equal-time to null plane dynamics, and the commutator (5). We have of course commutation outside the light cone, but a singularity on the light cone. It is this singularity that has to be dealt with.

We consider explicitly the Fermi field; the Bose field can be treated similarly. The free field anticommutator is

$$\left.\begin{aligned}
&\{\psi(x),\ \bar{\psi}(y)\} = -i\,(\gamma_\mu\partial_\mu - m)\,\Delta(x-y)\\
&\Delta(x) = -\frac{\varepsilon(x_0)}{2\pi}\left(\delta(x^2) - \frac{m^2}{2}\,\frac{J_1(\sqrt{-m^2x^2})}{\sqrt{-m^2x^2}}\,\theta(-x^2)\right)\\
&\varepsilon(x_0) = x_0/|x_0| \qquad\qquad \theta(-x^2) = \tfrac{1}{2} + \tfrac{1}{2}\varepsilon(-x^2)
\end{aligned}\right\} \tag{37'}$$

The Bessel function can be expanded in powers of its argument, beginning with the first power.

The important point here is that the light cone singularities of the anticommutator (37)′ are either independent of, or proportional to a finite integral power of, the mass $m$. They can be removed therefore by Pauli-Villars regularization. The field $\psi(m)$ with mass $m$ can be replaced in the interaction Hamiltonian by a combination of two fields

$$\psi(m) \rightarrow \psi(m) + i\,\psi(m+M) \tag{38}$$

And one can further regularize by again substituting in the <u>regularized</u> interaction

$$\begin{aligned} \psi(m) &\to \psi(m) + i\,\psi(m+M) \\ \psi(m+M) &\to \psi(m+M) + i\,\psi(m+M+M) \end{aligned} \tag{39}$$

And so on, to obtain the desired degree of regularization. For a sufficiently regulated theory, the formal reasoning leading to (4) is presumably sound, and the problem is shifted to that of subsequently taking the limit $M \to \infty$.

The questions of regularization, renormalization, and the limit $M \to \infty$ arise in ordinary perturbation theory in connection with closed loop diagrams. What is striking here is that the regularization is important even for tree diagrams. Consider for example the interaction [part of the vector interaction (22)]

$$\begin{aligned} g\,\psi^{+} A_0 \psi &= g\,\hat{\psi}_g^{+} A_0 \hat{\psi}_g \\ &+ g\,\hat{\psi}_g^{+} \frac{m\sigma_3 - \vec{\sigma}_\perp \cdot \vec{p}_\perp'}{\eta'} A_0 \frac{m\sigma_3 - \vec{\sigma}_\perp \cdot \vec{p}_\perp}{\eta} \hat{\psi}_g \end{aligned} \tag{40}$$

This is explicitly dependent on $m$, and in the regulated version there are therefore terms which increase with the mass $(m+M)$ of the regulating field. The propagator for $\hat{\psi}_g$ is

$$P = \frac{1/2}{p_0 - \frac{1}{2}\left(\eta + \frac{m^2 + \vec{p}_\perp^{\,2}}{\eta}\right)} \tag{41}$$

and

$$\lim_{m \to \infty} m^2 P = -\eta \tag{42}$$

Thus effectively we have to add to the unregulated interaction (40) a whole series of contact terms. In order $g^2$ (40) is augmented by, from the regulation of the second term,

$$g^2\,\hat{\psi}_g^{+} \frac{m\sigma_3 - \vec{\sigma}_\perp \cdot \vec{p}_\perp'}{\eta'} A_0 \frac{1}{\eta_I} A_0 \frac{m\sigma_3 - \vec{\sigma}_\perp \cdot \vec{p}_\perp}{\eta} \hat{\psi}_g$$

where the reciprocal $(\eta_I^{-1})$ of the intermediate $\eta$ can be interpreted as an integral operator.

Such contact terms are familiar from other approaches to null plane dynamics [1),4)]. In the external potential case [4)] we dismissed them as relativistic corrections of second order (fourth order relative to $mc^2$). That is again so here, at least if we restore the suppressed factors c and count them.

Is it sensible to separate these particular regularization effects from the renormalization programme as a whole ? Is it sensible to speak of non-relativistic approximation, or to order terms by powers of c, when interactions are strong and loop diagrams important ? And so on.

## APPENDIX

Notation : Central European Standard [3].

### Plane waves

The plane wave Dirac spinors $u(p,s)$ with $p_o = \pm\sqrt{m^2+\vec{p}^2}$, satisfy

$$p_o u = (\beta m + \vec{\alpha}\cdot\vec{p})\, u \tag{A.1}$$

$$\sum_{\varepsilon, s} u_i\, u_j^* = |2p_o|\, \delta_{ij} \tag{A.2}$$

where the summation is for given $\vec{p}$ over $\epsilon = p_o/|p_o| = \pm 1$ and two spin states $s$.

The corresponding Foldy-Wouthuysen spinors $w$ satisfy

$$p_o w = \beta \sqrt{m^2 + \vec{p}^2}\; w \tag{A.3}$$

$$\sum_{\varepsilon, s} w_i\, w_j^* = |2p_o|\, \delta_{ij} \tag{A.4}$$

and are related to the $u$ by

$$u = \frac{m + \sqrt{m^2+\vec{p}^2} - i\vec{\gamma}\cdot\vec{p}}{\sqrt{(m+\sqrt{m^2+\vec{p}^2})^2 + \vec{p}^2}}\; w \tag{A.5}$$

where

$$i\beta\vec{\gamma} = \vec{\alpha} \tag{A.6}$$

### Good representation

$$\left.\begin{array}{lll} \alpha_3 = \begin{pmatrix} 1 & \\ & -1 \end{pmatrix} & \beta = \begin{pmatrix} & \sigma_3 \\ \sigma_3 & \end{pmatrix} & \vec{\alpha}_\perp = \begin{pmatrix} & -\vec{\sigma}_\perp \\ -\vec{\sigma}_\perp & \end{pmatrix} \\ \sigma_1 = \begin{pmatrix} & \sigma_1\sigma_3 \\ \sigma_1\sigma_3 & \end{pmatrix} & \sigma_2 = \begin{pmatrix} & \sigma_2\sigma_3 \\ \sigma_2\sigma_3 & \end{pmatrix} & \sigma_3 = \begin{pmatrix} \sigma_3 & \\ & \sigma_3 \end{pmatrix} \end{array}\right\} \tag{A.7}$$

## Good and bad components

The 4-spinors can be separated into pairs of good $(\alpha_3 = +1)$ and bad $(\alpha_3 = -1)$ 2-spinors. From (A.1)-(A.3)

$$u_b = \eta^{-1} \left( m\sigma_3 - \vec{\sigma}_\perp \cdot \vec{p}_\perp \right) u_g \tag{A.8}$$

$$w_b = p_0^{-1} \sqrt{m^2 + \vec{p}^{\,2}} \; w_g \tag{A.9}$$

where

$$\eta = p_0 + p_3 \tag{A.10}$$

For given $\vec{p}$ and $\epsilon(= p_0/|p_0| = \eta/|\eta|)$

$$\sum_s u_{gi} \, u^*_{gj} = \delta_{ij} \, |\eta| \tag{A.11}$$

$$\sum_s w_{gi} \, w^*_{gj} = \delta_{ij} \, |p_0| \tag{A.12}$$

## Free field operators

We associate with each plane wave an operator $a$, which is a destruction or creation operator for $\epsilon = \pm 1$, respectively. Normalizing covariantly

$$\left\{ a(p,s), a^+(p',s') \right\} = |2p_0| (2\pi)^3 \, \delta^3(\vec{p} - \vec{p}\,') \, \delta_{ss'} \, \delta_{\epsilon\epsilon'} \tag{A.13}$$

The ordinary Dirac field operator is then

$$\psi = \sum_{\epsilon,s} \int \frac{d^3\vec{p}}{(2\pi)^3 |2p_0|} \, a(p,s) \, u(p,s) \, e^{i\vec{p}\cdot\vec{x} - ip_0 t} \tag{A.14}$$

The Foldy-Wouthuysen field is

$$\phi = \sum_{\epsilon,s} \int \frac{d^3\vec{p}}{(2\pi)^3 |2p_0|} \, a(p,s) \, w(p,s) \, e^{i\vec{p}\cdot\vec{x} - ip_0 t} \tag{A.15}$$

The null plane Dirac field is

$$\hat{\psi}_g = \sum_{\epsilon,s} \int \frac{d^3\vec{p}}{(2\pi)^3|2p_0|} \, a(p,s)\, u_g(p,s)\, e^{i\vec{p}\cdot\vec{x} - ip_0(t-z)} \tag{A.16}$$

The null plane Foldy-Wouthuysen field is

$$\hat{\phi}_g = \sum_{\epsilon,s} \int \frac{d^3\vec{p}}{(2\pi)^3|2p_0|} \, a(p,s) \left|\frac{\eta}{p_0}\right| w_g(p,s)\, e^{i\vec{p}\cdot\vec{x} - ip_0(t-z)} \tag{A.17}$$

Note that

$$\vec{p}\cdot\vec{x} - p_0(t-z) = \eta z + \vec{p}_\perp\cdot\vec{x}_\perp - p_0 t \tag{A.18}$$

$$\sum_\epsilon \int \frac{d^3\vec{p}}{(2\pi)^3|2p_0|} = \int \frac{d\eta\, d^2\vec{p}_\perp}{(2\pi)^3|2\eta|} \tag{A.19}$$

$$\sqrt{m^2+\vec{p}^{\,2}} = \tfrac{1}{2}\eta + (m^2+\vec{p}_\perp^{\,2})/(2\eta) \tag{A.20}$$

From the basic anticommutator (A.13) and the completeness relations (A.2), (A.4), (A.11) and (A.12), we obtain the canonical anticommutators

$$\{\psi_i(t,\vec{x}),\, \psi_j^+(t,\vec{y})\} = \delta_{ij}\,\delta(\vec{x}-\vec{y}) \tag{A.21}$$

$$\{\phi_i(t,\vec{x}),\, \phi_j^+(t,\vec{y})\} = \delta_{ij}\,\delta(\vec{x}-\vec{y}) \tag{A.22}$$

$$2\,\{\hat{\psi}_{gi}(t,\vec{x}),\, \hat{\psi}_{gj}^+(t,\vec{y})\} = \delta_{ij}\,\delta(\vec{x}-\vec{y}) \tag{A.23}$$

$$2\,\{\hat{\phi}_{gi}(t,\vec{x}),\, \hat{\phi}_{gj}^+(t,\vec{y})\} = \delta_{ij}\,\delta(\vec{x}-\vec{y}) \tag{A.24}$$

Free field equations

$$i\,\frac{\partial\psi}{\partial t} = (\vec{\alpha}\cdot\vec{p} + \beta m)\,\psi \tag{A.25}$$

$$i \frac{\partial \phi}{\partial t} = \beta \sqrt{m^2 + \vec{p}^2} \; \phi \tag{A.26}$$

$$i \frac{\partial \hat{\psi}_g}{\partial t} = \left( \frac{\eta}{2} + \frac{m^2 + \vec{p}_\perp^{\,2}}{2\eta} \right) \hat{\psi}_g \tag{A.27}$$

$$i \frac{\partial \hat{\phi}_g}{\partial t} = \left( \frac{\eta}{2} + \frac{m^2 + \vec{p}_\perp^{\,2}}{2\eta} \right) \hat{\phi}_g \tag{A.28}$$

From these free field equations, and from the Heisenberg equation of motion

$$i \dot{o} = [o, H] \tag{A.29}$$

can be inferred several forms for the free Hamiltonian

$$H_o = \sum_{\epsilon, s} \int \frac{d^3 \vec{p}}{(2\pi)^3 |2 p_o|} \; p_o \, a^+(p,s) \, a(p,s) \tag{A.30}$$

$$= \int d^4x \; \delta(t) \; \psi^+ (\vec{\alpha} \cdot \vec{p} + \beta m) \, \psi \tag{A.31}$$

$$= \int d^4x \; \delta(t) \; \phi^+ \beta \sqrt{m^2 + \vec{p}^2} \; \phi \tag{A.32}$$

$$= 2 \int d^4x \; \delta(t) \; \hat{\psi}_g^+ \left( \frac{\eta}{2} + \frac{m^2 + \vec{p}_\perp^{\,2}}{2\eta} \right) \hat{\psi}_g \tag{A.33}$$

$$= 2 \int d^4x \; \delta(t) \; \hat{\phi}_g^+ \left( \frac{\eta}{2} + \frac{m^2 + \vec{p}_\perp^{\,2}}{2\eta} \right) \phi_g \tag{A.34}$$

## Foldy-Wouthuysen transformation

From (A.5)

$$\psi = \frac{m + \sqrt{m^2 + \vec{p}^2} - i \vec{\gamma} \cdot \vec{p}}{\sqrt{(m + \sqrt{m^2 + \vec{p}^2})^2 + \vec{p}^2}} \; \phi \tag{A.35}$$

$$= e^{-iS} \, \phi \, e^{iS}$$

$$S = - \int d^4x \, \delta(t) \, \psi^+ \arctan \frac{\vec{\gamma} \cdot \vec{p}}{m + \sqrt{m^2 + \vec{p}^2}} \, \psi \qquad (A.36)$$

Second Melosh transformation

From (A.35)

$$\hat{\psi}_g = \frac{m + \sqrt{\eta^2} + \sigma_3 \vec{\sigma}_\perp \cdot \vec{p}_\perp}{\sqrt{(m + \sqrt{\eta^2})^2 + \vec{p}_\perp^{\,2}}} \, \hat{\Phi}_g \qquad (A.37)$$

$$= e^{-iY} \, \hat{\Phi}_g \, e^{iY}$$

$$Y = -2 \int d^4x \, \delta(t) \, \hat{\psi}_g^+ \arctan \frac{i \sigma_3 \vec{\sigma}_\perp \cdot \vec{p}_\perp}{m + \sqrt{\eta^2}} \, \hat{\psi}_g \qquad (A.38)$$

## REFERENCES


1) Reviewed by :
J. Kogut and L. Susskind, Phys.Reports 8C (1973);
Ph. Meyer, Cargèse Lectures (1972);
R. Jackiw, Springer Tracts in Modern Physics 62 (1972).

2) A similar programme has been initiated by :
R. Carlitz and Wu-Ki Tung, Chicago preprint (1975).

3) J.S. Bell, Acta Phys. Austriaca, Suppl. 13, 395 (1974).

4) J.S. Bell and H. Ruegg, Nuclear Phys. B93, 12 (1975).

5) J.S. Bell and H. Ruegg, CERN preprint TH.2095 (1975).

6) G. Morpurgo, in Theory and Phenomenology in Particle Physics, Ed., A. Zichichi, Erice 1968; Academic Press (1970). See also the introductory remarks in this volume.

7) H.J. Melosh, Thesis, California Institute of Technology, Pasadena (February, 1973). See also Ref. 3).

# CURRENT AND CONSTITUENT QUARKS: THEORY AND PRACTICE

A.J.G. Hey

Physics Department, Southampton University

## Abstract

The present status of $SU(6)_W$ phenomenology is surveyed in detail. After a discussion of SU(6) multiplets in the baryon and meson spectra, some ideas underlying the construction of $SU(6)_W$ decay models are reviewed. In particular, the approach to $SU(6)_W$ invoking the Melosh transformation between current and constituent quarks is described, and also, more briefly, explicit harmonic oscillator quark model calculations. The relative successes of the predictions of these two approaches for both baryon and meson decays are discussed at length. Some interesting suggestions emerge concerning the missing meson multiplets.

## Contents

## 1. The Constituent Quark Model Spectrum

The constituent quark model[1] visualizes hadrons as being built (in some sense at any rate) from basic spin $\frac{1}{2}$ quarks

$$q \sim (u,d,s) \times (\uparrow,\downarrow) \sim 6$$

This leads naturally to the idea that SU(6) multiplets should be identifiable in the resonance spectrum. For Baryons, the allowed representations are just those of three quarks (qqq)

$$6 \otimes 6 \otimes 6 = 56 \oplus 70 \oplus 70 \oplus 20$$

The total angular momentum $\underline{J}$ of the resonance is decomposed into quark spin $\underline{S}$ and orbital angular momentum $\underline{L}$

$$\underline{J} = \underline{L} + \underline{S}$$

and we expect the lowest state to have L=0 and positive parity[2]. A count of the spin and unitary spin states of the nucleon and delta multiplets leads to the well-known result

$$\left.\begin{array}{ll} \Delta: & {}^4 10 = 40 \\ N: & {}^2 8 = 16 \end{array}\right\} \quad \text{56 spin and unitary spin states}$$

Thus, these states are assigned to a $\{56, L^P=0^+\}$ representation of SU(6) $\otimes$ O(3). If one now looks in detail at the negative parity states one finds the remarkable result that all negative parity resonances below about 2 GeV in mass can be assigned to a $\{70,1^-\}$ representation of SU(6) $\otimes$ O(3)[3]. This is not a trivial prediction: the existence of a well established 'extra' state would have caused the model grave difficulties. Fig. 1 shows the present status of the Y=0 and Y=1 states of the $\{70,1^-\}$. It should be remembered, however, that we are for the moment ignoring any mixing between states of the same spin and parity: this is meant as a 'best guess' at the dominant quark spin of each resonance. Furthermore, there are some missing Y* states and also some peculiarities concerning the $\Sigma(1660)$ $3/2^-$ state[4]. (However, it is probably fair to say that this situation reflects the relative 'youth' of Y* phase shift analyses, compared to the well established N* analyses). On the whole, the agreement is impressive. Proceeding through the Data Tables one can identify a $\{56,2^+\}$ and indications of a possible $\{56,4^+\}$ multiplet. The positive parity Roper resonance P11 N(1430), which is well in the midst of the negative parity 70-plet, is conventionally assigned to a "radial

| | | | |
|---|---|---|---|
| | | $^{4}\underset{\sim}{8}$ | |
| 5/2⁻ | N*(1670) | Σ(1675) | Λ(1830) |
| 3/2⁻ | N*(1710 | Σ(1940) | Λ(  ) |
| 1/2⁻ | N*(1660) | Σ(  ) | Λ(1670) |
| | | $^{2}\underset{\sim}{8}$ | |
| 3/2⁻ | N*(1520) | Σ(1660) | Λ(1690) |
| 1/2⁻ | N*(1510) | Σ(  ) | Λ(  ) |
| | | $^{2}\underset{\sim}{10}$ | |
| 3/2⁻ | Δ(1700) | Σ(1580) | |
| 1/2⁻ | Δ(1610) | Σ(1780) | |
| | | $^{2}\underset{\sim}{1}$ | |
| 3/2⁻ | | | Λ(1520) |
| 1/2⁻ | | | Λ(1805) |

Fig. 1. Status of Y=0 and Y=1 States of the $\{70,1^-\}$

excitation" of the ground state - a $\{56,0^+\}_R$ multiplet. We shall say more about this later.

From the baryon spectrum it seems that it does indeed make sense to classify resonances into SU(6) supermultiplets. How does the spectrum compare with that of the harmonic oscillator shell model pioneered by Greenberg and Dalitz[3)]? Fig. 2 shows the average multiplet masses predicted by Horgan in his recent paper[6)]. There seems to be little if any evidence for the many states of the $\{70,0^+\}$ and $\{70,2^+\}$ multiplets predicted to be roughly degenerate with the $\{56,2^+\}$[3)]. The possible absence of these harmonic oscillator multiplets at low masses may mean that it is time to break away from the straight-jacket of harmonic oscillator models. Having said this, for the mesons we shall find that there is little else to fall back on!

The allowed SU(6) representations for mesons are those of a quark-antiquark system ($q\bar{q}$)

$$6 \otimes \bar{6} = 35 \oplus 1$$

and the $J^{PC}$ quantum numbers are the same as those of the $N\bar{N}$ system

$$\underline{J} = \underline{L} + \underline{S}$$
$$P = (-1)^{L+1}$$
$$C = (-1)^{L+S}$$

The ground state 35 + 1 is expected to have L=0 and negative parity: the pion and rho nonets are assigned to the $\{35{+}1:\ L^P{=}0^-\}$ multiplet of SU(6) x O(3).

$$\left.\begin{array}{l} J^{PC} = 0^{-+}\ :\ "\pi"\ \ {}^1 8 + {}^1 1 = 9 \\ J^{PC} = 1^{--}\ :\ "\rho"\ \ {}^3 8 + {}^3 1 = 27 \end{array}\right\}\quad \begin{array}{l}\text{35+1 Spin and} \\ \text{Unitary Spin States}\end{array}$$

At the L=1 level the picture is still very confused. States with the predicted spin and parity

$$L^P = 1^+ : J^{PC} = 1^{+-},\ 2^{++},\ 1^{++},\ 0^{++}$$

have been observed but there are many missing states. Table I shows a fairly realistic view of the present sorry state of affairs. What about higher excited states? The harmonic oscillator quark model predicts six $J^{PC}$ multiplets at the N=2 level:

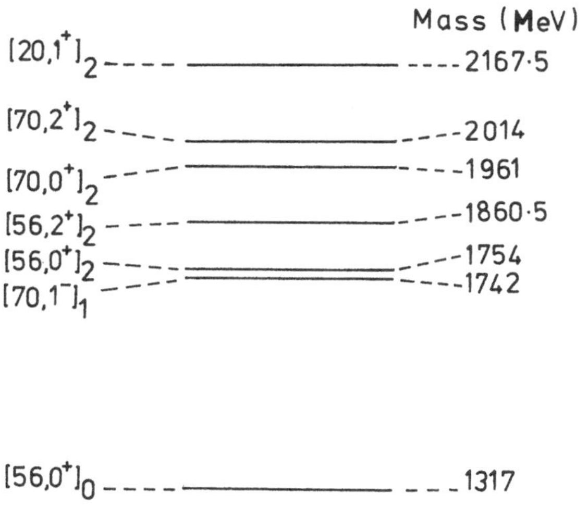


Fig. 2. Mass spectrum of Harmonic Oscillator Quark Model

Table 1 : Status of N=0 and N=1 Meson multiplets

| Oscillator Quantum Numbers N | L | S | $J^{PC}$ | $\underline{8}$ I=1 | $\underline{8}$ I=½ | $\underline{8}$ and I=0 | $\underline{1}$ I=0 | Mixing |
|---|---|---|---|---|---|---|---|---|
| 0 | 0 | 0 | $0^{-+}$ | $\pi$ | K | $\eta$ | $\eta'$ | U |
| | | 1 | $1^{--}$ | $\rho$ | K* | $\omega$ | $\phi$ | M |
| 1 | 1 | 0 | $1^{+-}$ | B | $Q_B$? | [$B_\eta$] | [$B_\eta'$] | [U] |
| | | 1 | $2^{++}$ | $A_2$ | K** | f | f' | M |
| | | | $1^{++}$ | $A_1$? | $Q_A$? | D? | [D'] | [M] |
| | | | $0^{++}$ | $\delta$? | K? | $\varepsilon$? | S*? | M? |

Notation: States that are in some doubt are marked?; states for which there is little or no evidence are enclosed [ ] . The mixing is indicated by U (unmixed) or M ('magic' or ideal mixing), whichever is nearer to the physical situation.

N=2 Mesons

$$\begin{array}{llll} & & 3^{--} & g \\ L=2 : 2^{-+} & A_3 & 2^{--} & X \\ & & 1^{--} & \rho^{**} \\ & & & \\ L=0 : 0^{-+} & \pi^* & 1^{--} & \rho^* \end{array}$$

Of these six multiplets, there are no trustworthy candidates for at least three of them. The N=2 L=0 states are the meson "radial excitations" - of crucial importance for quark interpretations of the new resonances found at Brookhaven and SLAC. The score sheet for these N=2 multiplets is given in Table 2. A new $J^{PC} = 4^{++}$ state has recently been reported[7)8)]. It has been seen in the $K^+K^-$ and $\pi^o\ \pi^o$ spectra with a mass around 2025 MeV and a width of about 200 MeV. This is presumably to be assigned to an N=3 L=3 quark model multiplet.

It is clear then that the meson spectrum and SU(6) multiplets are in much worse shape than the Baryons. However, this should not surprise us since, for several reasons, meson spectroscopy is much more difficult. For example, even though Baryon resonances may be directly observed in formation experiments, only three or four 'bumps' are seen in the cross section. The vast majority of resonant states have emerged from painstaking phase shift analysis. Typically, for stable $\pi N$ phase shift solutions, roughly $10^4$ events are needed for the integrated differential cross section, $\int\frac{d\sigma}{dt}\,dt$, at each energy. Meson states, on the other hand, may only be studied in production experiments (apart from the obvious exceptions of $e^+e^-$ and $\bar{p}p$ annihilation reactions). Meson phase shift analysis thus requires a dynamical production model - usually single pion exchange or Reggeon exchange - to remove the unwanted final state Baryon. There are still further difficulties. Since quantum number restrictions are much more stringent for meson systems, to observe unnatural parity meson resonances one needs to study reactions involving three final state mesons. Unfortunately the analysis of such reactions is complicated and difficult. The theoretical problems arise because one must use a dynamical model to perform a $J^P$ analysis, and the isobar model which is used does not satisfy two body unitarity in the final state[9)]. (The corrections are "probably small" but it is disturbing that we don't know for certain).

An Aside on the $A_1$ Crisis

Isobar analyses of $\pi N \to \pi\pi N$ in the Baryon resonance region seem to be a spectacular success for $SU(6)_W$ models - as we shall see in section 3. However, isobar model analyses of the three pion system in diffractive $\pi N \to \pi\pi\pi N$ reactions have obstinately refused to show any hint of resonance structure in the $1^+$ channel.

Table 2 - Present status of N=2 multiplets (notation as in Table 1)

| N | L | S | $J^{PC}$ | $\underline{8}$ I=1 | $\underline{8}$ I=½ | $\underline{8}$ I=0 | and $\underline{1}$ I=0 | Mixing |
|---|---|---|---|---|---|---|---|---|
| 2 | 0 | 0 | $0^{-+}$ | [$\pi^*$] | [$K_{\pi^*}$] | [$\eta^*$] | [$\eta^{*\prime}$] | [U] |
| | | 1 | $1^{--}$ | $\rho^*$? | [$K_{\rho^*}$] | [$\omega^*$] | [$\phi^*$] | [M] |
| 2 | 2 | 0 | $2^{-+}$ | $A_3$? | L? | [$\eta_{A_3}$] | [$\eta'_{A_3}$] | [U] |
| | | 1 | $3^{--}$ | g | $K_g$ | $\omega_g$ | [$\phi_g$] | M? |
| | | | $2^{--}$ | [X] | [$K_x$] | [$\omega_x$] | [$\phi_x$] | [M] |
| | | | $1^{--}$ | [$\rho^{**}$] | [$K_{\rho^{**}}$] | [$\omega^{**}$] | [$\phi^{**}$] | [M] |

The Ascoli analyses[10] seem to show that most if not all of the $1^+$ signal comes from what is euphemistically known as the Deck "background"[11]! To avoid this non-resonant background in the $1^+$ channel, one has to look at non-diffractive reactions and this is hard experimentally[12]. Even here, the preliminary indications for $A_1$ hunters are not encouraging[13]. However, since it is difficult to build quark models without an $A_1$ resonance somewhere in the mass range 1.0 to 1.4 GeV and I, for one, am unhappy with proposals that do without it[14], one is forced to put down its non-observance to inadequacies of the isobar model, theoretical ambiguities lack of statistics, etc, but... If the newly discovered intermediate states found at DESY and SLAC do turn out to be the P-states of charmonium, it is embarassingly true that this would be our strongest evidence for the existence of an $A_1$! In section 4, however, we shall see that the quark model suggests several interesting new places to look for the $A_1$.
End of aside.

In summary there are genuine reasons to worry about the uniqueness and validity of multiparticle phase shift analyses. Even in the case of $\pi N \rightarrow \pi\pi N$, at each energy there are four kinematic variables so that a phase shift analysis of comparable accuracy to the $\pi N$ elastic analyses would require many more than $10^4$ events per energy. What then is to be done? In the absence of any neutral phenomenological procedure it is sensible to look into the possibility of using theoretical models to guide our view of the meson world. Theoretical models with an explicit dynamical content can provide a guide as to what behaviour to expect, where to look and which parameterization to use in fitting data. Clearly one must choose a model with a successful 'track record'. In section 3 we will describe how $SU(6)_W$ models motivated via the Melosh transformation and, to a lesser extent, the harmonic oscillator quark model successfully correlate large amounts of data for baryon resonance decays. Then in section 4, we shall apply similar technqiues to the N=1 and N=2 meson multiplets and see an interesting picture emerge. Before this, in section 2, we shall briefly review the ingredients of both algebraic Melosh $SU(6)_W$ decay models and explicit harmonic oscillator quark models for resonance transitions.

## 2. SU(6) Models for Resonance Decays

### 2.1. Algebraic $SU(6)_W$ Models à la Melosh

From the experimentally observed approximate $SU(6)_W$ multiplet structure we deduce the existence of a set of $SU(6)_W$ generators $\hat{W}_\alpha$, that transform a one-particle state to another resonance within the same $SU(6)_W$ multiplet

$$\hat{W}_\alpha \, |1\rangle \sim |1'\rangle$$

However, we have no knowledge as to whether the $\hat{W}_\alpha$ operators are physically measurable. On the other hand, from current algebra we know of the existence of a set of null-plane charges $\hat{F}_\alpha$, that also generate an $SU(6)_W$ algebra and that are the integrals of physically measurable currents[15]. Further, these null-plane charges annihilate the vacuum

$$\hat{F}_\alpha |0\rangle = 0$$

thereby avoiding Coleman's theorems and the problems associated with the pair states produced by non-conserved equal-time charges. These charges may therefore be used to classify states into approximate $SU(6)_W$ multiplets. It is natural to suppose that $\hat{W}_\alpha$ and $\hat{F}_\alpha$ are related: Gell-Mann suggested the relation

$$\hat{W} = V\hat{F}V^{-1}$$

where V is a unitary transformation[16]. Setting V=1 and identifying the two sets of generators, however, leads to well-known phenomenological difficulties, and also theoretical contradictions[17]. We must accept therefore the fact that $V \neq 1$. The problem of <u>deriving</u> a realistic form for V is very difficult and in these lectures we shall not discuss the various attempts[18]. Rather, we shall take a phenomenological point of view and regard Melosh's essentially free quark model form for V[19] as merely the simplest phenomenological expression of the fact that V cannot be unity.

For example, under the $SU(6)_W$ of currents the pion charge $Q_\pi$ transforms as a 35 with W-spin one and z-component zero

$$(Q_\pi)_{\hat{F}} \sim 35\ (\pi\text{: } W=1\, W_z=0)$$

Under the constituent $SU(6)_W$, Melosh's transformation suggests that $Q_\pi$ remains in a 35 but transforms as the sum of two representations

$$(Q_\pi)_{\hat{W}} \sim \alpha\{35\text{: } W=1\, W_z=0;\ L_z=0\}$$

$$+\ \beta\{35;\ W=1\ W_z=\pm 1;\ L_z=\mp 1\}$$

or in an obvious shorthand

$$(Q_\pi)_{\hat{W}} \sim \alpha\{35;\ '\pi'\} + \beta\{35;\ 'A_1'\}$$

In principle, there could be many other terms present - such as exotic 405 representations - but this is the most general form within 35 representations. The assumption clearly reflects our hope that exotic pieces are relatively unimportant - which seems to be supported by the successful confrontation of the resulting $\pi$ decay model with data. To apply these ideas about $SU(6)_W$ symmetry to pionic decays $A \to B+\pi$ one must approximate the matrix element by the single particle matrix element of $Q_\pi$ using PCAC[20][21]

$$\langle B\pi|A\rangle \underset{PCAC}{\sim} \langle B|Q_\pi|A\rangle$$

The SU(6) algebraic structure may now be calculated using standard Clebsch Gordan techniques, and the predictions are couched in terms of a small number of unknown reduced matrix elements which are the parameters of this model.

By making similar assumptions concerning the transformation properties of the electromagnetic current components under $SU(6)_{W,\text{constituents}}$ one can obtain $SU(6)_W$ models for photon[22][23] and rho decays of resonances [24][25].

2.2. Explicit Harmonic Oscillator Quark Models

The explicit quark model calculations [1] in contrast to the Melosh approach, construct a detailed dynamics at the quark level, usually for a harmonic oscillator potential between quarks. Decay transitions are characterized by one-quark operators which automatically restrict the operators to 35 represenatations. For example, for pion decays, the pion-emission operator $H_\pi$ for $L=1 \to L=0$ transitions has the simple form

$$H_\pi = A\sigma_z L_z + B\sigma_{\pm} L_{\mp}$$

where the $L_{z,\pm}$ operators are orbital excitation operators and $\sigma_{z,\pm}$ quark spin operators. The similarity with $(Q_\pi)_{\hat{W}}$ is obvious and both models in fact yield the same SU(6) algebraic structure[26][27]. Explicit quark models make further, more specific predictions. Firstly, A and B are explicit functions of the resonance masses involved in the decay and consequently these models have a much more detailed way of incorporating SU(6) mass splitting effects than the Melosh approach[28] Secondly, harmonic oscillator models predict the relative magnitudes of the reduced matrix elements A and B (corresponding to $\alpha$ and $\beta$ of the Melosh model) for different SU(6) multiplet transitions. There are therefore two levels of predictions, first the SU(6) algebraic structure and then, if this is successful, the more detailed intermultiplet relations of more explicit models. As to the ways of incorporating mass splitting effects we shall make some remarks about the two approaches in section 4 for the mesons.

The explicit quark models construct similar single-quark transition operators for real and virtual photon transitions. However, in most models used to date, the most general 35 structure is not assumed - a more specific operator is used[29]. Again, it seems sensible, given a reasonable amount of data, to test the most general 35 operator structure against data before restricting oneself to more specific parameterizations.

In the next section we review the main features of algebraic $SU(6)_W$ fits to Baryon decays and comment on the fate of the more detailed predictions of other models. For mesons, however, there is little or no data and we must resort to explicit models to give us a possible qualitative overall picture of meson decays.

## 3. Baryon Decays and $SU(6)_W$ Phenomenology

### 3.1. Pionic Decays

The $SU(6)_W$ model based on the Melosh transformation and PCAC was outlined in section 2.1 - it is clearly much more predictive than SU(3). For decays of the Y=1 states of the $\{70,1^-\}$ to the $\{56,0^+\}$, the 21 independent SU(3) coupling constants are related in terms of known SU(6) coupling coefficients and only two unknown parameters - which we can choose as S-wave and D-wave amplitudes. Similarly the $\{56,2^+\}$, and $\{56,0+\}_R$ decays to the ground state are described in terms of P and F amplitudes, and a P' amplitude, respectively. Unfortunately, this simple picture is somewhat marred by symmetry breaking effects. Before we can apply the model to data, some choices must be made as to how obvious SU(3) symmetry breaking effects - such as the mass differences of $\pi$, K and $\eta$ - are to be incorporated when the model is extended to include Y* decays. Furthermore, since our SU(6) multiplets are by no means mass degenerate, the question of barrier factors is important for numerical agreement. For example, Gilman, Kugler and Meshkov[20] retain in the amplitude a factor $(M_A{}^2 - M_B{}^2)$ arising from PCAC, and use the same barrier factor for decays with different partial waves. Recent phenomenological analysis[30] shows that some dependence on the 3-momentum of the decay p which varies with the partial wave angular momentum $\ell$, such as $p^\ell$, seems to give best agreement. This was also found by Faiman and Plane in their earlier $SU(6)_W$ fit[31]. Finally, since we have N* and Y* states of the same spin and parity in the $\{70,1^-\}$ and $\{56,2^+\}$ multiplets, there is the possibility of mixing between the pure SU(6) states. In our analysis[30], we found that a good fit to the $\{56,2^+\}$ was obtained without mixing but that the fit to the $\{70,1^-\}$ was improved with mixing. One must be careful in interpreting these results - $SU(6)_W$ with PCAC is not exact. A good measure of its validity may be gauged from the $5/2^-$ $\Lambda$* D05(1830) which must be unmixed and is predicted to decouple from $\bar{K}N$. However, this resonance is clearly seen in $\bar{K}N\to\Sigma\pi$ and accountes for a $\chi^2$ contribution of 36 in our fit. Nevertheless this $\chi^2$ contribution is somewhat misleading since the elasticity of this resonance is less than 10% -

i.e. small but non-zero - to be compared with the zero prediction of $SU(6)_W$. We take the view that this is the level of accuracy of the model - until we have a theory of symmetry breaking effects. Alternatively, one could believe the model predictions to better than 10% and try to "explain" the discrepancy by intermultiplet mixing say, with a $\Lambda^*$ from a possible $\{70,3^-\}$. However, since SU(6) representation mixing is also a symmetry breaking effect this seems a somewhat circular procedure. In our fit we found that to a reasonable approximation the states could be regarded as unmixed - apart from singlet-octet mixing among the $\Lambda^*$'s. Even in this approximation of zero mixing, the SU(6) fits were not much worse than SU(3) fits (see below). However, the Y* mixing matrices from our fit and the SU(6) fit made by Horgan[6] to resonance masses do not show good agreement. Furthermore, Jones et al.[32] attempted to fit the masses and decays simultaneously and were unable to find any consistent set of mixing matrices[33]! From a Melosh approach, I am not sure that it is possible to define an $SU(6)_W$ symmetry limit for the mass operator. I am therefore not clear about the relative status of $SU(6)_W$ fits to decay rates and masses. It is clearly an interesting problem!

After all these caveats, let me try to convince you that these $SU(6)_W$ fits are spectacularly successful. Consider the two general types of $SU(6)_W$ predictions - amplitude signs[34] and magnitudes.

(1) Amplitude Signs

For $N_1^*$ and $N_2^*$ in the same SU(6) multiplet the model predicts the relative signs of inelastic amplitudes i.e. $A_1/A_2$ where

$$A_1 = \text{amplitude for } \pi N \to N_1^* \to \pi\Delta$$

$$A_2 = \text{amplitude for } \pi N \to N_2^* \to \pi\Delta$$

A quick check of the number of non-trivial sign predictions for the $\{70,1^-\}$ and $\{56,2^+\}$ decays yields a total of 16 in agreement with the $SU(6)_W$ model and at present no obvious discrepancies.

(2) Magnitudes

The detailed fits to experimental amplitudes $\sqrt{xx'}$ are generally quite good (see refs. 25 and 30) and here we demonstrate this in two ways

(i) In order to assess the significance of the $SU(6)_W$ Clebsch-Gordan coefficients, we have attempted fits to the data using random numbers (normalized between $\pm$ 1) in place of them in the theory predictions. (We of course keep the SU(3) coefficients). The results are shown in fig. 3: the SU(6) coefficients gave a $\chi^2$ of 34 compared to the best "random fit" with a $\chi^2$ of 94.

(ii) We compare the results of our $SU(6)_W$ analysis with more familiar SU(3) analyses. In Fig. 4(a) we attempt to give an idea of the relative success of these two types of fits by plotting the ratio

$$R = g^{SU(3)}_{Expt} \;/\; g^{SU(3)}_{Model}$$

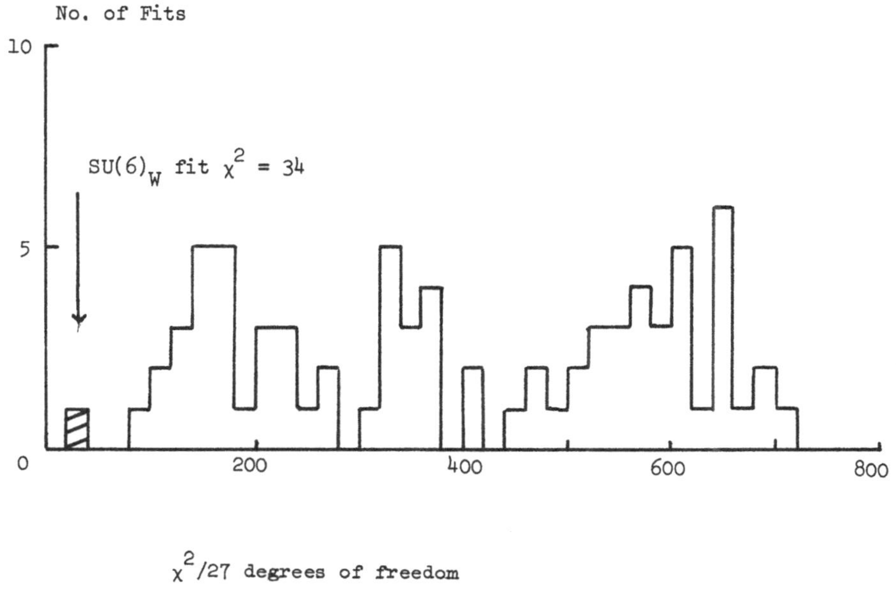


Fig. 3. Random Number Fits to $\{56,2^+\}$ $\pi$ Decays

for members of both the $\{70,1^-\}$ and $\{56,2^+\}$ multiplets. The coupling constants $g^{SU(3)}_{Expt}$ are taken from independent analyses[35] of N*, Σ* and Λ* resonances in the various $J^P$ multiplets. The $g^{SU(3)}_{Model}$ are calculated using the best values of the SU(6) parameters (S,D,P and F) together with the predicted SU(6) F/D ratio (ignoring mixing). From the agreement in Fig. 4a where the results for the various ratios are seen to cluster around one, it is clear that our SU(6) fit (of 2 parameters to 55 data points for the $\{70,1^-\}$ and 2 parameters to the 27 data points of the $\{56,2^+\}$) fares quite well compared with the much less ambitious SU(3) fits. The errors shown are those quoted by the SU(3) analyses and are very probably underestimated. But in Fig. 4b all is not well for the $\frac{1}{2}^-$ states. Our SU(3) predictions however, were calculated <u>as if</u> the three $\frac{1}{2}^-$ resonances used in the SU(3) analysis were actually the $^2 8$ and $^2 1$ SU(6) states. A look at the SU(6) fit[30] gives the reason for the failure: the states used in the SU(3) analysis, the N*(1520), Λ*(1670) and Σ*(1740) are best classified in our analysis as predominantly $^2 8$, $^4 8$ and $^2 10$ respectively! Thus the SU(3) analysis of these states is innappropriate.

In summary we must conclude that there is strong evidence for $SU(6)_W$ structure in these pion decays. The amplitude signs may be categorized as follows:

For the $\{70,1^-\}$: "Anti-$SU(6)_W$"
(signs as when $\alpha_{70}(L_z=0)$ term is absent)

For the $\{56,2^+\}$: "$SU(6)_W$-like"
(signs as when $\beta_{56}(L_z=\pm 1)$ term is absent)

The result "$SU(6)_W$-like" for the $\{56,2^+\}$ is determined solely from the N*F15(1680) sign in the Berkely-SLAC πΔ analysis[36], but is supported by the ΛF05(1820) sign in the CHS Σ*π analysis[37]. Some of the more explicit quark models require both the 70 and the 56 multiplets to be "anti-$SU(6)_W$", but it appears impossible to dismiss the evidence. In the framework of the isobar model analyses of πΔ the F15 signs are well determined and cannot be tampered with![38]. Thus the $^3P_o$ model in the form used by le Yaouanc[39], which incorporates quark orbital wavefunctions for the states, predicts the anti-$SU(6)_W$ sign for both 70 and 56 in disagreement with experiment. Similarly, Feynman, Kislinger and Ravndal's version of the harmonic oscillator quark model[40] is in contradiction with experiment for the same reason[41].

Finally, some comments on possible radial excitations - $\{56,0^+\}_R$. These multiplets contain $^2 8$ and $^4 10$ states and there are two candidates for $^2 8$ P11 nucleon states - the NP11(1430) 'Roper' resonance, and an NP11(1750). When fitting the $\{70,1^-\}$ multiplet, after allowing for the freedom for an overall sign in inelastic amplitudes - one for πΔ amplitudes from Berkeley-SLAC, one for πΣ* amplitudes, etc. - and for the relative signs of S and D, we obtained ten well-determined correct sign predictions. However, for the $\{56,2^+\}$, $\{56,4^+\}$ and $\{56,0^+\}_R$ fits we are even more tightly constrained since there is no longer a sign ambiguity remaining for

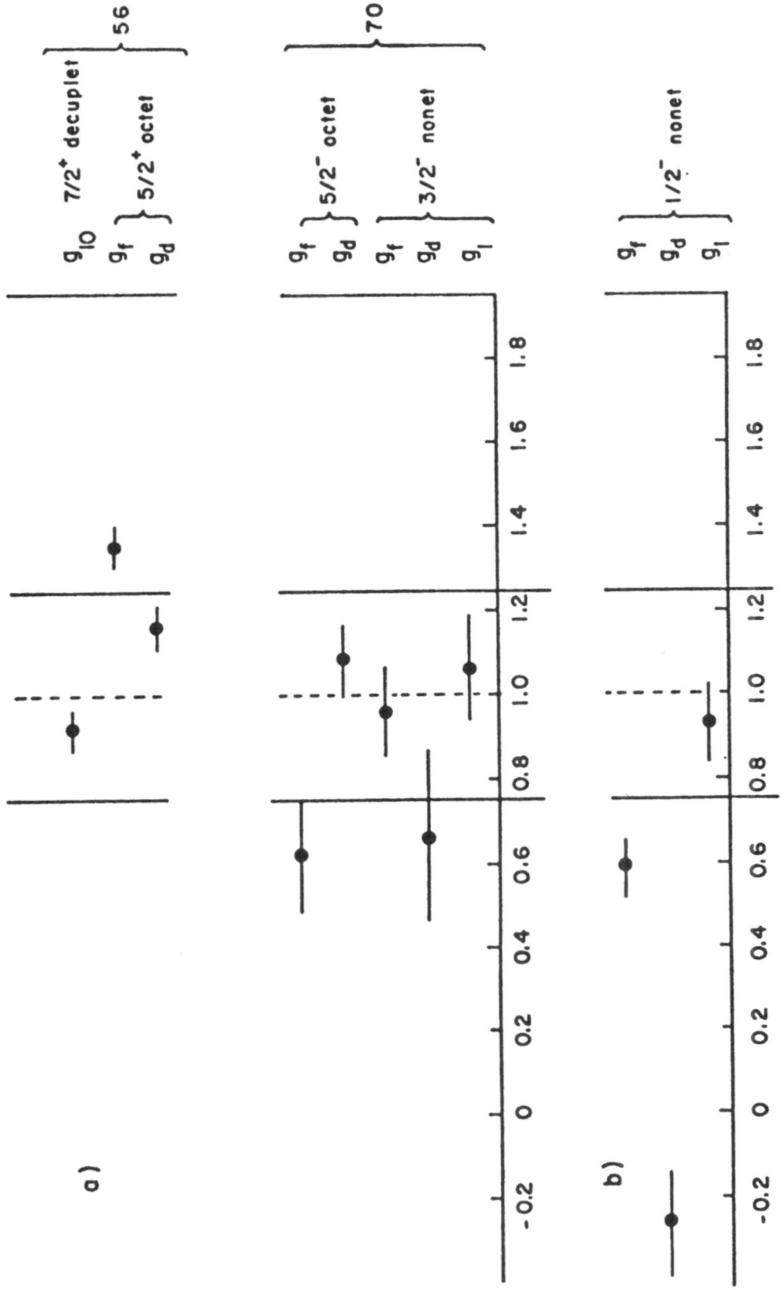


Fig. 4. Comparison of SU(3) couplings vs. SU(6) Predictions

the inelastic amplitudes. Thus, determining the $\pi\Delta$ amplitudes in the $\{70,1^-\}$ analysis, enables us to say with certainty that the P11(1430) and the P11(1750) cannot be classified in 70 multiplets. Possible assignments in a $\{70,0^+\}$ or $\{70,2^+\}$ yield the opposite sign to experiment and so if these P11's are to be classified as bona-fide quark resonances in SU(6) multiplets[42] then they must both belong to 56's. In the absence of positive parity L=1 multiplets, the only assignment is $\{56,0^+\}$. These assignments for the P11(1430) and the P11(1750) are supported by an independent SU(6) analysis of photoproduction data (see later). Nevertheless, a quantitative fit to the Roper multiplet including the P33 (1690) is not very satisfactory. Since the P11(1750) must now be ruled out as a candidate for a $\{70,0^+\}$ multiplet[43], the explicit energy level calculations of conventional harmonic oscillator quark models must be in some trouble.

3.2. Photon Transitions

Resonance photoproduction may be subjected to a similar algebraic $SU(6)_W$ analysis. For real photons there are at most two independent helicity amplitudes for resonance excitation: the amplitudes for an helicity ½ proton to absorb $\lambda = \pm 1$ photons.

$$F_\pm \sim \langle N^* | J_\pm | N \rangle$$

where $J_\pm$ represents the electromagnetic current operator corresponding to $\lambda = \pm 1$ photons. To formulate an $SU(6)_W$ model we must know the transformation properties of $J_\pm$ under $SU(6)_W$ constituents. The Melosh transformation suggests the following structure[22),23)]

$$(J_\pm)_{\hat{W}} \sim A\left[35;{}^{W=0}_{W_z=0}\ ;\ L_z=\pm 1\right] + B\left[35;{}^{W=1}_{W_z=\pm 1};\ L_z=0\right]$$
$$+ C\left[35;{}^{W=1}_{W_z=0};\ L_z=\pm 1\right] + D\left[35;{}^{W=1}_{W_z=\mp 1};\ L_z=\pm 2\right]$$

The $SU(6)_W$ predictions for $\gamma N \to N^*$ are now just straightforward Clebsch-Gordan algebra. Putting these results together with the predictions for $N^* \to N\pi$ decays leads to amplitudes for $\gamma N \to \pi N$. For the $\{70,1^-\}$ and $\{56,2^+\}$ multiplets the results of our fit to the photoproduction analysis of Walker and Metcalf[44] are shown in Table 3. The conclusions may be summarized as

For the $\{70,1^-\}$: $A_{70} \neq C_{70}$; $C_{70} \neq 0$

For the $\{56,2^+\}$: $A_{56} \sim C_{56}$; $C_{56} \neq 0$; $D_{56} \neq 0$

Table 3 : Photoproduction Fits to the $\{\underset{\sim}{70},1^-\}$ and $\{\underset{\sim}{56},2^+\}$

3a $\{\underset{\sim}{70}, 1^-\}$

| | $\chi^2$ | $A_{70}$ | $B_{70}$ | $C_{70}$ |
|---|---|---|---|---|
| 1. Melosh $SU(6)_W$ Fit<br>$A_{70}$, $B_{70}$, $C_{70}$ free | 25.4 | 8.3 | 2.2 | 4.2 |
| 2. Quark Model Fit<br>$C_{70} = 0$ | 73.4 | 8.9 | 3.9 | 0 |
| 3. $^3P_0$ Model Fit<br>$A_{70} = C_{70}$ | 47.3 | 6.6 | 0.6 | 6.6 |

3b $\{\underset{\sim}{56}, 2^+\}$

| | $\chi^2$ | $A_{56}$ | $B_{56}$ | $C_{56}$ | $D_{56}$ |
|---|---|---|---|---|---|
| 1. Melosh $SU(6)_W$ Fit<br>$A_{56}$, $B_{56}$, $C_{56}$, $D_{56}$ free | 10.7 | -6.7 | -1.1 | -6.8 | 4.6 |
| 2. Quark Model Fit<br>$C_{56} = D_{56} = 0$ | 65.2 | -4.8 | -1.8 | 0 | 0 |
| 3. $^3P_0$ Model Fit<br>$A_{56} = C_{56}$; $D_{56} = 0$ | 32.8 | -5.9 | -1.1 | -5.9 | 0 |

These conclusions agree very well with those of Balscock and Rosner[45] who have recently performed a critical appraisal of photoproduction analyses in terms of multipoles, and consider three recent analyses[44][46][47]. All lead to the conclusion that the 'C terms' are definitely present and important[48]. This result is sufficient to rule out the simplest versions of harmonic oscillator quark models which have C = D = 0 for all multiplets.[29][40]
The $^3Po$ model[49] which predicts A = C and D = 0 fares little better : in the $(56,2^+)$ case it seems that $D_{56}$ is definitely present. However, the $B_{56}$ term is rather ill determined and the advent of new data especially on photoproduction amplitude signs for the P13 resonance are forecast to cause trouble for these $SU(6)_W$ fits and a re-analysis is needed. Still, as Moorhouse[50] noted at Palermo, "we must still regard the sign of this photoproduction amplitude as being only dubiously determined."

Carlitz and Weyers[51] exploit the idea of an expansion of a non-local Melosh-type transformation in powers of a fundamental length $a = 1/m$. This leads to predictions of the relative importance of the various $SU(6)_{W,\text{constituent}}$ terms in the transformed current operator. In particular, it leads to the prediction that the pionic transitions of the $(70,1^-)$ should be "anti-$SU(6)_W$" whereas those of the $(56,2^+)$ should be "$SU(6)_W$-like" - in agreement with experiment. Applied to photon transitions, the approach predicts for the $(70,1^-)$ $A_{70}$ and $C_{70}$ should dominate, but for the $(56,2^+)$ $B_{56}$ and $D_{56}$ should be the dominant contributions. Although this is consistent with $(70,1^-)$ transitions, and there does seem to be a need for a significant contribution from $D_{56}$, both $A_{56}$ and $C_{56}$ are also important. So at present the status of these ideas is unclear.

A last few words on $(56,0^+)$ excitation. The $SU(6)_W$ structure implies the following predictions for P33 and P11 amplitudes (in the notation of Walker and Metcalf)

$$\text{P33} \qquad \frac{A_{1+}}{B_{1+}} = -\frac{1}{2}$$

$$\text{P11} \qquad \frac{A^P_{1-}}{A^n_{1-}} = +\frac{3}{2}$$

From their fit we find

$$\text{P33 (1238)} \qquad \frac{A_{1+}}{B_{1+}} = -0.48 \pm .02$$

$$\text{P11 (1430)} \qquad \frac{A^P_{1-}}{A^n_{1-}} = +1.62 \pm 1.4$$

$$\text{P11 (1650)} \qquad \frac{A^{p}_{1-}}{A^{n}_{1-}} = +1.42 \pm 1.3$$

Within the very large errors, this suggests that both P11's probably lie in $(56,0^+)$ multiplets rather more than in possible $(\underline{70},0^+)$ or $(\underline{70},2^+)$ multiplets (which predict 3 or zero for the ratio respectively). However it is clear that the pion and photon transitions of these possible $(56,0^+)$'s need further study. Nevertheless, it is worth remarking that this algebraic $SU(6)_W$ model is a distinct improvement on explicit quark models which are unable[44) 46)] to predict the signs of the P11(1430) amplitudes correctly.

We shall not discuss the extension of these $SU(6)_W$ models to virtual (spacelike) photon transitions : these will be discussed by Frank Close at this workshop.

3.3. Rho Decays

Amplitudes for the rho decays may only be treated within the framework of an $SU(6)_W$ scheme motivated by the Melosh transformation by relating them to matrix elements of the electromagnetic current. Therefore we extrapolate the algebraic structure of $\lambda = \pm 1$ photon transitions from $q^2 = 0$ to $q^2 = M_\rho^{\,2}$ and assume that the isovector portion approximates the $SU(6)_W$ structure of transverse rho meson decays. In this approach to SU(6), since the $\pi$ and $\rho$ decays of baryons involve assumptions about different current operators, their decay parameters need not be related in any obvious manner even though $\pi$ and $\rho$ are in the same SU(6) spectrum multiplet. Longitudinal rho amplitudes $F_o$, may be related to matrix elements of $J^+ \sim J^o + J^3$ , the 'good' component of the electromagnetic current.

$$F_o \sim \langle A|J^+|B\rangle$$

Under $SU(6)_{W,\text{constituent}}$ $J^+$ is assumed to transform as[24) 25)]

$$(J^+)_{\hat{W}} \sim a^o(35{:}W{=}0\ L_z{=}0) + a^1(35{:}W{=}1W_z{=}\pm 1\ ;\ L_z{=}\mp 1)$$

Thus in this algebraic approach there are many free parameters to fit the very modest number of $N\rho$ amplitudes determined in the Berkeley-SLAC analysis[36)]. Restricting the freedom by making a vector dominance extrapolation from the photoproduction values leads to an acceptable fit, with no helicity zero rhos, although a significantly better fit is obtained if the longitudinal rho amplitude $a^o_{56}$ is included [25)]. The results of this latter fit are shown in Table 4.

Table 4 : $N\rho$ Decays of the $\{70,1^-\}$ and $\{56,2^+\}$

| Resonance | Wave | Expt. Value | Prediction |
|---|---|---|---|
| S11(∿1500) | $SS11\rho_1$ | −.12 ± .08 | −.15 |
| D13(∿1520) | $DS13\rho_3$ | −.32 ± .10 | −.28 |
| P13(∿1730) | $PP13\rho_1$ | +.35 ± .10 | +.17 |
| F15(∿1680) | $FP15\rho_3$ | −.27 ± .10 | −.39 |
| F35(∿1860) | $FP35\rho_3$ | +.28 ± .08 | +.04 |
| The resonances below were not used in the fit since there are experimental problems in determining the sign of the resonant amplitudes. The signs in parentheses are those determined by a T matrix fit. | | | |
| S31(∿1610) | $SS31\rho_1$ | ±.18 ± (+) | +.36 |
| D33(∿1700) | $DS33\rho_3$ | ±.20 ± (−) | +.17 |
| S11(∿1660) | $SS11\rho_1$ | ±.23 ± (−) | −.19 |
| D13(∿1700) | $DS13\rho_3$ | ∿ 0 ± (+) | −.02 |
| F37(∿1920) | $FF37\rho_3$ | ±.18 ± (+) | −.05 |

This is in contradiction with Faiman's suggestion[52] that the best fit is obtained with no longitudinal rho amplitudes.

In conclusion, we must warn against taking numerical values for this $\rho$ decay analysis too seriously. There are not enough well determined amplitudes to allow more than preliminary comparisons to be made, but it is clear that the model does not lead to glaring sign disagreements at present. The more explicit quark models do, however, appear to predict some wrong signs[53], and this, when coupled with the indications from the baryon spectrum and from the pion and photon transitions, would appear to indicate that these models only give at best a qualitative picture of the baryon spectrum and decays.

## 4. Meson Decays and $SU(6)_W$ Predictions

### 4.1 Quark Model Predictions

For the baryons there exists data on $N\pi$, $\Delta\pi$, $N\gamma$ and $N\rho$ decays of resonances and the SU(6) models are quite well constrained. For the mesons we do not yet have the resonances let alone their decay systematics! In this situation, the algebraic approach of the Melosh transformation is not very predictive since it requires a modicum of reasonable data to determine the reduced matrix elements of the model. In order to obtain some predictive power we must use some explicit quark model. Extrapolating from the quark model's qualitative success for the baryons, this should give a good qualitative guide for experimental meson spectroscopy.

There are many varieties of quark models available for specific calculations, generally differing in only fairly minor ways[1]. The calculations presented here are based on the covariant harmonic oscillator model developed by Feynman, Kislinger and Ravndal (FKR)[40] who performed similar calculations for the N=0 and N=1 SU(6) supermultiplets. In this model, for decays to non-ground state mesons the pion emission operator $H_\pi$ has the form[27]

$$\begin{aligned} H_\pi = G e_\pi e^{\bar{\rho}\tilde{a}_o^+} \{ & (d\rho)\sigma_z + (-dx\rho^2)\sigma_z a_z \\ & + (dx\rho^2)\sigma_z \tilde{a}_o^+ + b\sigma_z a_z - b\sqrt{2}\sigma_+ a_+ \\ & + b\sqrt{2}\sigma_- a_- + \gamma b \sigma_z \tilde{a}_o^+ \\ & - b\sqrt{2}\sigma_- a_+^+ + b\sqrt{2}\sigma_+ a_-^+ \} \, e^{-\rho a_z} \end{aligned}$$

where $\rho, \bar{\rho}, d, x, \gamma$ and b are explicit functions of the particle masses, and $a_\mu$ ($a_\mu^+$) are the annihilation (creation) operators of orbital excitations.

More details of the calculations may be found in Ref. 27 - here we shall only make the following comments :

(1) Unphysical timelike excitations have been excluded from the spectrum by the covariant condition

$$P^{\mu} a_{\mu} | \text{Physical state; } P> = 0$$

This accounts for much of the complication of the form for $H_{\pi}$ appropriate for decays to moving L-excited states.

(2) The factor G has the form

$$G = \exp\left[ - \frac{Q^2}{\Omega} \frac{M_A{}^2}{M_A{}^2 + M_B{}^2} \right]$$

where $\Omega \sim 1$ GeV/c$^2$ is the Regge slope parameter. This is a heuristic attenuation factor introduced by FKR in order to make a rough correction for the distortion of unitarity caused by excluding time-like states. The form of the factor is physically quite plausible and furthermore, in our applications it does not play a crucial role, since it rarely produces more than a factor of 2 suppression[54).]

(3) The model has some limitations. For example, we must always work in a narrow resonance, quasi-two-body final state approximation

i.e. $M^* \rightarrow M' + \pi$
$\quad\quad\quad \hookrightarrow M + \pi$ etc.

Thus for $4\pi$ states, the modes "$\rho\varepsilon$" and "$\rho\rho$" which are sometimes quoted, are not calculated by this approach. We shall make some comments on these "$\rho$-decays" later.

For numerical predictions, one must input the physical resonance masses. For the disputed states, these may be uncertain by several hundred MeV, and our quark model gives no reliable guide. The numerical predictions are therefore based on the specific (although somewhat arbitrary) set of masses detailed in Table 5. These are meant to be a "plausible" set of masses for the N=1 and N=2 multiplets. For the $A_1$ we have given two possibilities - light (1070 MeV) and heavy (1300 MeV). As a first approximation we have adopted a naive mixing rule for the I=0 states

quark spin 1 - magic mixing
quark spin 0 - unmixed

Further discussion and caveats, as well as consideration of other possibilities, are to be found in ref. 27.

Let us now look at the predictions of the model - beginning with the relatively well-known N=0 and N=1 states. On the whole there is good agreement - as can be seen from Table 6 and Fig. 5.

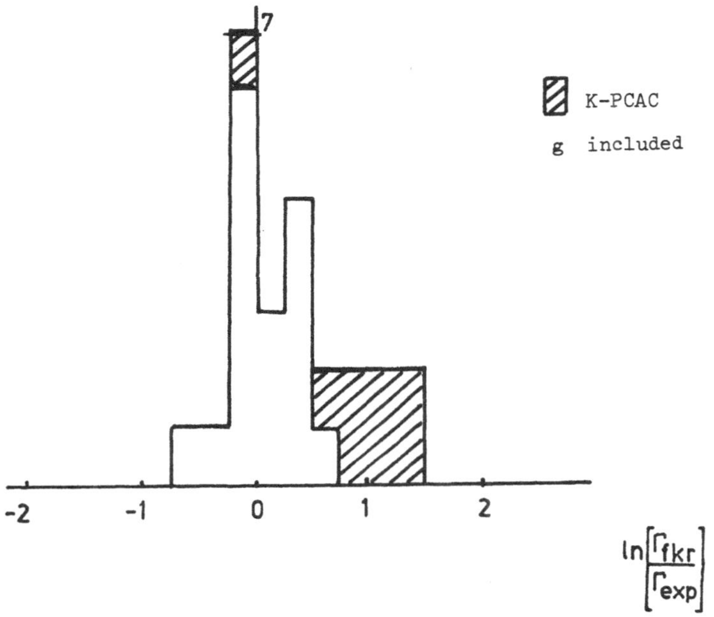


Fig. 5. 'Feynman' Plot of FKR vs. reliable meson data

Table 5 : "Standard" Mass Set for N=1 and N=2 Super-Multiplets

| Multiplet N,L | $J^{PC}$ | $^{2s+1}L_J$ | I=1 | I=1/2 | I=0 | I=0 | Mixing |
|---|---|---|---|---|---|---|---|
| N=1 L=1 | $1^{+-}$ | $^1P_1$ | B(1235) | $Q_B$(1300) | h(1250) | h'(1250) | U |
| | $2^{++}$ | $^3P_2$ | $A_2$(1310) | K**(1420) | f(1270) | f'(1514) | M |
| | $1^{++}$ | $^3P_2$ | $A_1$(1070) [1300] | $Q_A$(1240) | D(1285) | D'(1416) | M |
| | $0^{++}$ | $^3P_0$ | $\delta$(970) | $\kappa$(1100) | $\varepsilon$(1200) | S*(1000) | M |
| N=2 L=0 | $0^{-+}$ | $^1S_0$ | $\pi$*(1400) | $K_{\pi*}$(1500) | $\eta$*(1400) | $\eta'$*(1500) | U |
| | $1^{--}$ | $^3S_1$ | $\rho$*(1600) | $K_{\rho*}$(1700) | $\omega$*(1600) | $\phi$*(1800) | N |
| N=2 L=2 | $2^{-+}$ | $^1D_2$ | $A_3$(1640) | L(1740) | $\eta_{A_3}$(1700) | $\eta'_{A_3}$(1900) | U |
| | $3^{--}$ | $^3D_3$ | g(1680) | $K_g$(1800) | $\omega_g$(1660) | $\phi_g$(2000) | M |
| | $2^{--}$ | $^3D_2$ | X(1700) | $K_X$(1800) | $\omega_X$(1700) | $\phi_X$(1900) | M |
| | $1^{--}$ | $^3D_1$ | $\rho$**(1800) | $K_{\rho**}$(1900) | $\omega$**(1800) | $\phi$**(2000) | M |

U = Unmixed ($\eta,\eta'$)

M = Magic ($\omega,\phi$)

Table 6 : Quark Model Predictions for N=0 and N=1 Super-Multiplets

| Multiplet N,L | $J^{PC}$ | Resonance | Decay Mode | Partial Width Theory (MeV) | | Total Width Theory (MeV) | |
|---|---|---|---|---|---|---|---|
| $[\underset{\sim}{35}+\underset{\sim}{1}\ 0^-]$ | $1^{--}$ | $\rho$ | $\pi\pi$ | 158 | 150±10 | 158 | 150±10 |
| N=0 | | $K^*$ | $K\pi$ | 65 | 50±1 | 65 | 50±1 |
| L=0 | | $\phi$ | $K\bar{K}$ | 12 | 3±0 | 12 | 4±0 |
| $[\underset{\sim}{35}+\underset{\sim}{1},\ 1^+]$ | $1^{+-}$ | B | $\omega\pi$ | 84 | 120±20 | 100 [96] | 120±20 |
| N=1 | | | $A_1\pi$ | 4[o] | | | |
| L=1 | | | $\delta\pi$ | 12 | | | |
| | | $Q_B$ | $\rho K$ | 51 | | 110 | |
| | | | $K^*\pi$ | 44 | | | |
| | | | $\omega K$ | 14 | | | |
| | | | $\kappa\pi$ | 2 | | | |
| | | h | $\rho\pi$ | 95 | | | |
| | | h' | $\rho\pi$ | 189 | | | |

Table 6 (con't)

| Multiplet N, L | $J^{PC}$ | Resonance | Decay Mode | Partial Width Theory (MeV) | | Total Width Theory (MeV) | |
|---|---|---|---|---|---|---|---|
| $[35+1, 1^+]$ N=1 L=1 | $2^{++}$ | $A_2$ | $K\overline{K}$ | 19 | $5\pm1$ | 111 | $100\pm10$ |
| | | | $\eta\pi$ | 23 | $15\pm1$ | | |
| | | | $\eta'\pi$ | 2 | <1 | | |
| | | | $\rho\pi$ | 67 | $72\pm2$ | | |
| | | K** | $K\pi$ | 88 | $55\pm3$ | 128 | $100\pm10$ |
| | | | $\eta K$ | 5 | $2\pm2$ | | |
| | | | $\rho K$ | 8 | $9\pm2$ | | |
| | | | $K^*\pi$ | 24 | $30\pm3$ | | |
| | | | $\omega K$ | 2 | $4\pm2$ | | |
| | | | $Q_A\pi$ | 1 | | | |
| | | f | $\pi\pi$ | 241 | $150\pm20$ | 265 [255] | $170\pm30$ |
| | | | $K\overline{K}$ | 14 | $8\pm6$ | | |
| | | | $\eta\eta$ | <1 | | | |
| | | | $A_1\pi$ | 9[o] | | | |
| | | f' | $K\overline{K}$ | 105 | $40\pm10$ | 152 | $40\pm10$ |
| | | | $\eta\eta$ | 33 | | | |
| | | | $\eta'\eta$ | <1 | | | |
| | | | $K^*\overline{K}$ | 14 | | | |

Table 6 (con't)

| Multiplet N, L | $J^{PC}$ | Resonance | Decay Mode | Partial Width Theory (MeV) | | Total Width Theory (MeV) | |
|---|---|---|---|---|---|---|---|
| $[\underline{35}+\underline{1},1^+]$ N=1 L=1 | $1^{++}$ | $A_1$ | $\rho\pi$ | 151 [340] | | 151 [340] | |
| | | $Q_A$ | $K^*\pi$ | 59 | | | |
| | | | $\kappa\pi$ | $<1$ | | | |
| | | D | $A_1\pi$ | 20[0] | | 99[79] | $30\pm20$ |
| | | | $\delta\pi$ | 79 | | | |
| | | D' | $K^*\bar{K}$ | 211 | | 211 | |
| | $0^{++}$ | $\delta$ | $\eta\pi$ | 99 | | 99 | $50\pm20$ |
| | | $\kappa$ | $K\pi$ | 333 | | 362 | |
| | | | $\eta K$ | 29 | | | |
| | | $\varepsilon$ | $\pi\pi$ | 1230 | | 1549 | |
| | | | $K\bar{K}$ | 292 | | | |
| | | | $\eta\eta$ | 27 | | | |
| | | S* | $K\bar{K}$ | 178 | | 178 | $40\pm8$ |

NOTE: 1) Heavy $A_1$ predictions in parentheses [].

2) $K^*\bar{K}$ short-hand for $K^*\bar{K} + \bar{K}^*K$.

Note the following points.

(i) Whenever K PCAC has to be used the model yields rates a factor $\sim 2$ too large.

(ii) For $B \to \omega\pi$, the model also predicts the correct helicity amplitude ratio.

(iii) The well-established $2^{++}$ nonet is in very good agreement with the model.

(iv) For the $0^{++}$ nonet, the model predicts (with this choice of masses and mixing) a very broad "$\varepsilon$" with $\Gamma \sim 1500$ MeV! This we regard as a qualitative <u>success</u> of the model - since the model also predicts "normal" widths $\Gamma \sim 100$ MeV for other resonances. We now turn to the N=2 states. The model has no freedom - all parameters are already determined. The predictions are given in Table 7 and some comments are made below:

(a) <u>$3^{--}$ Nonet</u>

For this, the best known N=2 multiplet, there is very encouraging agreement. The predicted width of the g meson, $\Gamma_g \sim 180$ MeV, is in good agreement with experiment and should be contrasted with the width predicted for the "$\varepsilon$" $\Gamma_\varepsilon \sim 1500$ MeV.

(b) <u>$2^{-+}$ Nonet</u>

Reasonable widths and branching ratios are found for the $A_3$ and L, but comparison with experiment is difficult since these mesons are beset by the same interpretative difficulties as the $A_1$ and Q bumps.

(c) <u>$2^{--}$ Nonet</u>

The I=1 member decays mainly into $4\pi$, there are no two body modes, and the resonance is predicted to be very broad. There are similar predictions for other members of the nonet.

(d) <u>$0^{-+}$ Radial Excitation</u>

For $\pi^*$ at a mass of 1.4 GeV, the total width is predicted to be very narrow $\Gamma \sim 20$ MeV. For a mass of 1.6 GeV this increases to 170 MeV but with $\rho\pi$ still a few MeV, and decays into a $0^{++}$ meson plus $0^{-+}$ $(\pi,K,\eta)$ being dominant. One must remember that the widths into $\varepsilon\pi$ are calculated in the narrow resonance approximation: clearly for the $\varepsilon$ this is invalid. If one takes a popular $\varepsilon$ mass of around 700 MeV the $\varepsilon\pi$ decay width increases to hundreds of MeV. This gives some indication of the effect on the predictions of a very broad $\varepsilon$ resonance in the final state[55].

(e) <u>$1^{--}$ States</u>

The $\rho'(1600)$ must presumably be identified as the N=2 L=2 state or the N=2 L=0 state[56]. However, for a $\rho^{**}_{L=2}$ at a mass of 1600 MeV, the model predicts a sizeable $\pi\pi$ width of $\Gamma_{\pi\pi} \sim 100$ MeV - much too large to be accommodated by the data. Thus we identify the $\rho'(1600)$ with the $\rho^*_{L=0}$ radial excitation, in agreement with the folklore concerning its production in $e^+e^-$ collisions. Where then is the $\rho^{**}$? Three possibilities are given in Table 8.

The overall picture given by the quark model is thus extremely plausible. We shall see in section 4.3 what it has to say about the missing states. One element

Table 7 : Quark Model Predictions for the N=2 Super-Multiplet

| Multiplet N, L | $J^{PC}$ | Resonance | Decay Mode | Partial Width Theory (MeV) | Total Width Theory (MeV) |
|---|---|---|---|---|---|
| N=2 L=0 | $0^{-+}$ | $\pi^*_{1.4}$ | $\rho\pi$ | 6 | 20 |
| | | | $K^*\bar{K}$ | <1 | |
| | | | $\varepsilon\pi$ | 14 | |
| | | $K_{\pi^*}$ | $\rho K$ | 5 | 161 |
| | | | $K^*\pi$ | 2 | |
| | | | $\omega K$ | 2 | |
| | | | $K^*\eta$ | 2 | |
| | | | $\delta K$ | 79 | |
| | | | $\kappa\pi$ | 52 | |
| | | | $S^*K$ | 21 | |
| | | $\eta^*$ | $K^*\bar{K}$ | <1 | 96 |
| | | | $\delta\pi$ | 96 | |
| | | $\eta^{1*}$ | $A_2\pi$ | <1 | 240 |
| | | | $\delta\pi$ | 240 | |
| | $1^{--}$ | $\rho^*_{1.6}$ | $\pi\pi$ | 1 | |
| | | | $K\bar{K}$ | <1 | |
| | | | $K^*\bar{K}$ | 4 | |
| | | | $\omega\pi$ | <1 | |
| | | | $\rho\pi$ | 2 | |
| | | | $h\pi$ | 6 | |

Table 7 (con't)

| Multiplet N,L | $J^{PC}$ | Resonance | Decay Mode | Partial Width Theory (MeV) | Total Width Theory (MeV) |
|---|---|---|---|---|---|
| N=2 L=0 | $1^{--}$ | $\rho^*_{1.6}$ (con't) | $h'\pi$ | 13 | 190 [73] |
| | | | $A_2\pi$ | 4 | |
| | | | $A_1\pi$ | 160[43] | |
| | | $K_{\rho^*}$ | $K\pi$ | 1 | 144 [62] |
| | | | $\eta K$ | <1 | |
| | | | $\rho K$ | 1 | |
| | | | $K^*\pi$ | <1 | |
| | | | $\omega K$ | <1 | |
| | | | $\phi K$ | 1 | |
| | | | $K^*\eta$ | <1 | |
| | | | $Q_B\pi$ | 18 | |
| | | | $K^{**}\pi$ | 1 | |
| | | | $A_1K$ | 81 [0] | |
| | | | $Q_A\pi$ | 38 | |
| | | $\omega^*$ | $K\bar{K}$ | <1 | 69 |
| | | | $\rho\pi$ | <1 | |
| | | | $K^*\bar{K}$ | 4 | |
| | | | $\omega\eta$ | 2 | |
| | | | $B\pi$ | 64 | |
| | | $\phi^*$ | $K\bar{K}$ | <1 | 169 |
| | | | $K^*\bar{K}$ | 3 | |
| | | | $\phi\eta$ | 4 | |
| | | | $Q_B\bar{K}$ | 20 | |
| | | | $h\eta$ | 5 | |
| | | | $h'\eta$ | 2 | |
| | | | $Q_AK$ | 134 | |

Table 7 (con't)

| Multiplet N, L | $J^{PC}$ | Resonance | Decay Mode | Partial Width Theory (MeV) | Total Width Theory (MeV) |
|---|---|---|---|---|---|
| N=2 L=2 | $2^{-+}$ | $A_3$ | $\rho\pi$ | 90 | 160 |
| | | | $K^*\bar{K}$ | 17 | |
| | | | $f\pi$ | 46 | |
| | | | $D\pi$ | 2 | |
| | | | $D'\pi$ | <1 | |
| | | | $A_1\eta$ | <1 [0] | |
| | | | $\kappa\bar{K}$ | <1 | |
| | | | $\epsilon\pi$ | 5 | |
| | | | $\delta\eta$ | <1 | |
| | | L | $\rho K$ | 33 | 174 |
| | | | $K^*\pi$ | 29 | |
| | | | $\omega K$ | 11 | |
| | | | $\phi K$ | 6 | |
| | | | $K^*\eta$ | 18 | |
| | | | $K^{**}\pi$ | 22 | |
| | | | $A_1K$ | 8 [0] | |
| | | | $Q_A\pi$ | 15 | |
| | | | $\delta K$ | 11 | |
| | | | $\kappa\pi$ | 15 | |
| | | | $\epsilon K$ | <1 | |
| | | | $S^*K$ | 6 | |
| | | | $\kappa\eta$ | <1 | |
| | | $^{\eta}A_3$ | $K^*\bar{K}$ | 68 | 191 [156] |
| | | | $A_2\pi$ | 47 | |
| | | | $A_1\pi$ | 43 [8] | |
| | | | $\delta\pi$ | 29 | |
| | | | $\kappa\bar{K}$ | <1 | |
| | | | $S^*\eta$ | 4 | |

Table 7 (con't)

| Multiplet N, L | $J^{PC}$ | Resonance | Decay Mode | Partial Width Theory (MeV) | | Total Width Theory (MeV) | |
|---|---|---|---|---|---|---|---|
| N=2 L=2 | $2^{-+}$ | $\eta'_{A_3}$ | $A_2\pi$ | 173 | | 561 [459] | |
| | | | $f\eta$ | 25 | | | |
| | | | $A_1\pi$ | 182 [80] | | | |
| | | | $Q_A\bar{K}$ | 27 | | | |
| | | | $D\eta$ | <1 | | | |
| | | | $\delta\pi$ | 90 | | | |
| | | | $\kappa\bar{K}$ | 50 | | | |
| | | | $\varepsilon\eta$ | 1 | | | |
| | | | $S^*\eta$ | 12 | | | |
| | $3^{--}$ | g | $\pi\pi$ | 61 | 50±20 | 176 [136] | 180±30 |
| | | | $K\bar{K}$ | 11 | 4±4 | | |
| | | | $K^*\bar{K}$ | 3 | | | |
| | | | $\omega\pi$ | 19 | | | |
| | | | $\rho\eta$ | 2 | | | |
| | | | $h\pi$ | 5 | | | |
| | | | $h'\pi$ | 10 | | | |
| | | | $A_2\pi$ | 17 | | | |
| | | | $A_1\pi$ | 48 [8] | | | |
| | | Kg | $K\pi$ | 36 | | 150 [139] | 250±100 |
| | | | $\eta K$ | 23 | | | |
| | | | $\rho k$ | 13 | | | |
| | | | $K^*\pi$ | 16 | | | |
| | | | $\omega K$ | 4 | | | |
| | | | $\phi K$ | 1 | | | |
| | | | $K^*\eta$ | <1 | | | |
| | | | BK | 2 | | | |
| | | | $Q_B\pi$ | 22 | | | |
| | | | hK | <1 | | | |
| | | | h'K | <1 | | | |
| | | | $K^{**}\pi$ | 8 | | | |
| | | | fK | <1 | | | |
| | | | $A_1K$ | 10 [0] | | | |
| | | | $Q_A\pi$ | 14 | | | |
| | | | DK | <1 | | | |
| | | | $Q_A\eta$ | <1 | | | |

Table 7 (con't)

| Multiplet N, L | $J^{PC}$ | Resonance | Decay Node | Partial Width Theory (MeV) | | Total Width Theory (MeV) | |
|---|---|---|---|---|---|---|---|
| N=2 L=2 | $3^{--}$ | $\omega_g$ | $K\bar{K}$ | 10 | | 112 | 140±20 |
| | | | $\rho\pi$ | 54 | | | |
| | | | $K^*\bar{K}$ | 2 | | | |
| | | | $\omega\eta$ | 1 | | | |
| | | | $B\pi$ | 44 | | | |
| | | $\phi_g$ | $K\bar{K}$ | 61 | | 270 | |
| | | | $K^*\bar{K}$ | 57 | | | |
| | | | $\phi\eta$ | 16 | | | |
| | | | $Q_B\bar{K}$ | 54 | | | |
| | | | $h\eta$ | 25 | | | |
| | | | $h'\eta$ | 13 | | | |
| | | | $K^{**}\bar{K}$ | 8 | | | |
| | | | $Q_A\bar{K}$ | 38 | | | |
| | $2^{--}$ | X | $K^*\bar{K}$ | 31 | | 435 [333] | |
| | | | $\omega\pi$ | 57 | | | |
| | | | $\rho\eta$ | 16 | | | |
| | | | $h\pi$ | 2 | | | |
| | | | $h'\pi$ | 5 | | | |
| | | | $A_2\pi$ | 119 | | | |
| | | | $A_1\pi$ | 118 [16] | | | |
| | | | $\delta\pi$ | 84 | | | |
| | | | $\kappa\bar{K}$ | 2 | | | |
| | | $K_x$ | $\rho K$ | 47 | | 308 [283] | |
| | | | $K^*\pi$ | 38 | | | |
| | | | $\omega K$ | 15 | | | |
| | | | $\phi K$ | 12 | | | |
| | | | $K^*\eta$ | 3 | | | |
| | | | $BK$ | <1 | | | |
| | | | $Q_B\pi$ | 9 | | | |
| | | | $hK$ | <1 | | | |
| | | | $h'K$ | <1 | | | |
| | | | $K^{**}\pi$ | 36 | | | |

Table 7 (con't)

| Multiplet N, L | $J^{PC}$ | Resonance | Decay Mode | Partial Width Theory (MeV) | Total Width Theory (MeV) |
|---|---|---|---|---|---|
| N=2 L=2 | $2^{--}$ | $K_x$ (cont) | fK | 27 | |
| | | | $A_1K$ | 25 [0] | |
| | | | $Q_A\pi$ | 28 | |
| | | | DK | <1 | |
| | | | $Q_A\eta$ | <1 | |
| | | | δK | 21 | |
| | | | κπ | 30 | |
| | | | εK | <1 | |
| | | | S*K | 12 | |
| | | | κη | 4 | |
| | | $\omega_x$ | ρπ | 177 | 250 |
| | | | $K^*\bar{K}$ | 31 | |
| | | | ωη | 15 | |
| | | | Bπ | 25 | |
| | | | $\kappa\bar{K}$ | 2 | |
| | | $\phi_x$ | $K^*\bar{K}$ | 108 | 237 |
| | | | φη | 44 | |
| | | | $Q_B\bar{K}$ | 4 | |
| | | | hη | 2 | |
| | | | h'η | 1 | |
| | | | $Q_A\bar{K}$ | 29 | |
| | | | $\kappa\bar{K}$ | 49 | |
| | $1^{--}$ | $\rho^{**}_{i.s}$ | ππ | 104 | 608 [553] |
| | | | $K\bar{K}$ | 41 | |
| | | | $K^*\bar{K}$ | 21 | |
| | | | ωπ | 25 | |
| | | | ρη | 9 | |
| | | | $Q_B\bar{K}$ | 51 | |
| | | | hπ | 27 | |
| | | | h'π | 54 | |
| | | | Bη | 35 | |
| | | | $A_2\pi$ | 17 | |
| | | | $A_1\pi$ | 146[91] | |
| | | | $Q_A\bar{K}$ | | |

**Table** 7 (con't)

| Multiplet N, L | $J^{PC}$ | Resonance | Decay Mode | Partial Width Theory (MeV) | Total Width Theory (MeV) |
|---|---|---|---|---|---|
| N=2 L=2 | $1^{--}$ | $K_{\rho**}$ | $K\pi$ | 47 | 756 [744] |
| | | | $\eta K$ | 54 | |
| | | | $\rho K$ | 22 | |
| | | | $K^*\pi$ | 15 | |
| | | | $\omega K$ | 7 | |
| | | | $\phi K$ | 9 | |
| | | | $K^*\eta$ | 2 | |
| | | | $BK$ | 136 | |
| | | | $Q_B\pi$ | 50 | |
| | | | $hK$ | 15 | |
| | | | $h'K$ | 120 | |
| | | | $Q_B\eta$ | 13 | |
| | | | $A_2K$ | 1 | |
| | | | $K^{**}\pi$ | 6 | |
| | | | $fK$ | 1 | |
| | | | $A_1K$ | 75 [63] | |
| | | | $Q_A\pi$ | 42 | |
| | | | $DK$ | 65 | |
| | | | $Q_A\eta$ | 77 | |
| | | $\omega^{**}$ | $K\overline{K}$ | 41 | 530 |
| | | | $\rho\pi$ | 75 | |
| | | | $K^*\overline{K}$ | 21 | |
| | | | $\omega\eta$ | 9 | |
| | | | $B\pi$ | 246 | |
| | | | $Q_B\overline{K}$ | 51 | |
| | | | $h\eta$ | 3 | |
| | | | $h'\eta$ | 6 | |
| | | | $Q_A\overline{K}$ | 79 | |
| | | $\phi^{**}$ | $K\overline{K}$ | 64 | 873 |
| | | | $K^*\overline{K}$ | 48 | |
| | | | $\phi\eta$ | 27 | |
| | | | $Q_B\overline{K}$ | 314 | |
| | | | $h\eta$ | 164 | |
| | | | $h'\eta$ | 82 | |
| | | | $K^{**}\overline{K}$ | 2 | |
| | | | $Q_A\overline{K}$ | 171 | |

Table 8 : Predictions for a $\rho^{**}_{L=2}$ state

| $\rho^{**}$ Mass | $\Gamma\pi\pi$ | $\Gamma_{4\pi}$ | $\Gamma$ total |
|---|---|---|---|
| 1.2 | 120 | 4 | 130 |
| 1.6 | 130 | 160 | 400 |
| 1.8 | 100 | 200 | 550 |

The $4\pi$ mode comprises channels such as $\omega\pi$, $A_2\pi$, $A_1\pi$, etc. The predictions are for a light $A_1$.

that is lacking, however, is a reliable model for $\rho\varepsilon$ and $\rho\rho$ decays. An FKR model for $\rho$ emission - using vector dominance of the photon transition operator - is too simple and gives unreliable predictions. For example, in a decay such as $A_2 \to \rho\pi$ to obtain consistency for the two ways of performing the calculation - either using $H_\pi$ or $H_\rho$ - the "C term", omitted in the FKR model, is required[45]. However, while it is true we certainly lack a reliable model for calculating these decays, it is also clear that the experimental analyses of 4 body states are far from complete. The plethora of 4 body channels indicated by these quark model calculations suggest that one must think seriously about attempting a serious isobar-type analysis of these channels. Only then will we know whether channels such as $\rho\varepsilon$, $\rho\rho$, etc. are really important[57].

4.2. Melosh and the Quark Model

The form for $H_\pi$ for decays to excited states looks considerably more complicated than the form of the Melosh transformed pion charge. Nevertheless in the limit of SU(6) degenerate multiplets both lead to identical algebraic structures[26][27]. The correspondence for transitions to ground state mesons is easy to read off but for the L=2 → L=1 transitions the relation is more complicated. In the Melosh approach there are five independent reduced matrix elements for these transitions labelled by $a^{L_{z_i}\, L_{z_f}}$ where $L_{z_i}$ and $L_{z_f}$ are the Z-component of orbital angular momentum of the initial and final states, respectively. In the form for $H_\pi$ one can identify five types of operator products which correspond directly to the $a^{L_{z_i}\, L_{z_f}}$ amplitudes of the Melosh approach

$$a^{00} \sim a_o^+ a_z^2 \sigma_z$$

$$a^{01} \sim a_o^+ a_\pm a_z$$

$$a^{10} \sim a_\pm^+ a_z^2$$

$$a^{11} \sim \sigma_z a_z$$

$$a^{12} \sim \sigma_\pm a_\pm$$

The exact functions of d,ρ etc. multiplying these operator combinations are given in ref. 27. In the degenerate limit a fit of the Melosh approach to the FKR predictions gives perfect agreement confirming these theoretical arguments (and helping debug the program!) However, when the degeneracy is broken, the symmetry breaking is clearly taken into account in a complicated but explicit manner in the FKR model. In the Melosh model, however, all the symmetry breaking effects are described in the barrier factor function chosen for the partial wave amplitudes. It is therefore of interest to note that one can reproduce quite well the results of the FKR model for the non-degenerate situation with the Melosh model. Fits were performed[27] with various barrier factors and the best fit given by a Blatt and Weisskopf "$\rho^{\ell}$" form (suitably flattened with a range parameter R). It is interesting that there was less spread in the helicity amplitude ratios between FKR and the Melosh fit, than for the partial widths. However, we learn from this exercise that a simple 'average' mass splitting factor can reproduce quite well the effects of a very complicated mass splitting pattern. To distinguish between them in a statistically significant way would indeed be very difficult.

It is worth remarking that the Melosh model also fits the reliably measured meson decays quite well - See fig. 6.

4.3. Lost Mesons

It is clear that at the N=2 level there are many missing states but it is interesting to note that the quark model also suggests reasons why these "lost" mesons may be difficult to detect.

(1) The $2^{--}$ states are typical of one type of lost state. The X(I=1) has a predicted total width $\Gamma \sim 500$ MeV and decays predominantly into 4 body channels such as $A_2\pi$, $A_1\pi$ etc. The contrast with the g meson with $\Gamma \sim 160$ MeV and a 2 body $\pi\pi$ width of 60 MeV is striking. Such states would not have been found by present analyses.

(2) For the $\pi^*$ radial excitation the situation is different. The decays to ground state mesons exist but the predicted widths are very small and stable to mass variations[58] e.g. $\pi^* \to \rho\pi$ and $K_{\pi^*} \to K^*\pi$ are both only a few MeV. If the numbers are taken at face value, why haven't we seen the $\pi^*$ with its predicted total width of 20 MeV? There are two relevant comments. One is as remarked earlier that the $\varepsilon$ is not a narrow resonance and proper treatment of a broad $\varepsilon$ could lead to a very wide resonance mainly coupled to $\varepsilon\pi$[55]. The other is that the $\pi^* \to \rho\pi$ matrix element is small and this can also affect the production of this state. A quick look at the width predictions suggests that the $\pi^*$ is hard to produce with pion beams since '$\varepsilon$' exchange may be the dominant trajectory. However, one must also calculate couplings to mesons (like the f) which are mass-disallowed and then the large extrapolations needed to obtain the Reggeon couplings make definite statements about the leading trajectory rather difficult. One is on much safer ground in predicting that photoproduction of the $\pi^*$ by pion exchange will be very weak. Perhaps a good place to look for the $\pi^*$ is in $p\bar{p}$, but it is difficult to support this hope with reliable calculations.

(3) The $\rho^*$ radial excitation similarly has a very small predicted width to ground

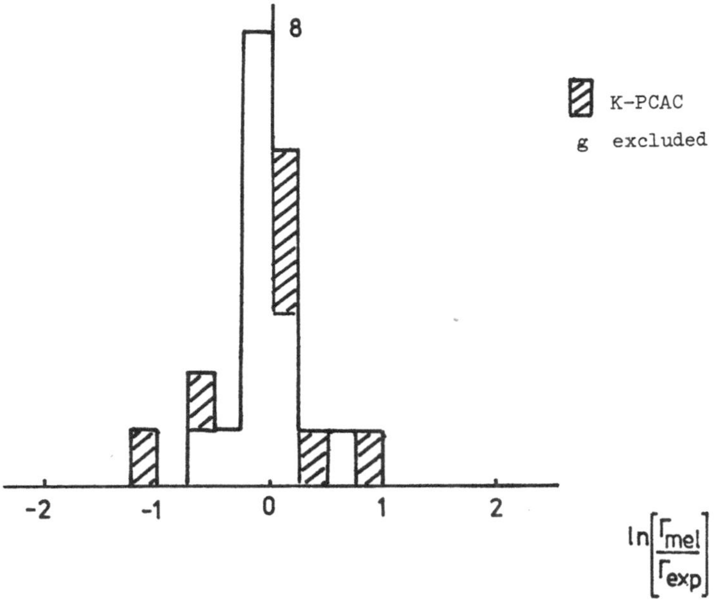


Fig. 6. 'Feynman' Plot of Melosh vs. reliable meson data

state mesons: $\rho^* \to \pi\pi \sim$ few MeV. Again a naive argument might suggest the leading trajectory for production by pions is the $A_1$, but again this must be hedged with caveats. Production of the $\rho^*$ in photoproduction by pion exchange is expected to be small (although, of course, one expects the $\rho^*$ to be diffractively produced by photons). Of interest in the context of Zweig's rule are the $\omega^*$ and the $\phi^*$. The model predicts they couple very weakly to the ground state mesons and so one must look for the $Q\bar{K}$ decay modes of the $\phi^*$ and the $B\pi$ decay mode of the $\omega^*$. They should also presumably be observable in $e^+e^-$ collisions and their relative photo-couplings are of great interest.

(4) The $\rho^{**}$ orbital excitation probably lies somewhere above the $\rho^*$ in mass. It has an appreciable $\pi\pi$ width and should be visible in higher mass $\pi\pi$ analyses.

The Quark model also suggests that 'cascade decays' may be useful to find some of the elusive N=1 mesons. The N=2 states include large branching fractions into N=1 states, and the decay of well-established higher states could provide a viable source of these "lost" mesons. A specific example is the g meson which, (for a light $A_1$), is predicted to have sizeable decay rates into $A_2\pi$ and $A_1\pi$. Similarly the $\rho^*$ also looks good for the $A_1$. Hopefully one can enhance the signal to noise ratio for the "lost" resonance, and certainly the background problems will be different from those of diffractive production. Nevertheless, for objects like the $A_1$ there seems no escaping a difficult 4 body analysis, be it $4\pi$ or $3\pi N$.

## 5. Concluding Remarks

For baryons, $SU(6)_W$ schemes have proved very successful both in the spectrum classification and for analysing $\pi,\gamma$ and $\rho$ transitions. It is clear though, that there are many problems of detail to resolve and in particular, the radial excitation pattern and the existence or non-existence of even parity 70 multiplets. However, the most urgent experimental questions concern meson spectroscopy. We must resolve the "$A_1$ crisis" (and related Q, $A_3$ and L crises) one way or the other and disentangle the resonance spectrum. The systematics of the meson radial excitations are obviously of great interest and have a bearing on theories of the new particles, as do the mixing patterns for the I=0 mesons for the different $J^P$ multiplets. When we have the spectrum. the detailed structure of $M^* \to M\pi$, $M^* \to M\gamma$ and $M^* \to M\rho$ transitions will be important in distinguishing between different SU(6) models. Indeed the photon decays of the ground state are already of interest and the recent measurement of $\phi \to \pi\gamma$ reported at the Stanford conference[59)] suggests that a new theoretical study of these decays would be rewarding. With a magically mixed $\phi$ the decay $\phi \to \pi\gamma$ is forbidden by Zweig's rule. Further investigation of Zweig's rule in processes such as $\phi^* \to \phi\pi\pi$ will also be very interesting - when we find the $\phi^*$. More knowledge of these "ordinary" mesons will undoubtedly aid our understanding of the new mesons discovered at Brookhaven, SLAC, and DESY.

The theoretical challenge is to incorporate these $SU(6)_W$ successes into a realistic dynamical model, presumably based on quarks. In the Melosh approach, this is the problem of deriving the form of the transformation in a realistic model. With the advent of gauge theories and models with quark confinement we may at last be making some progress - although many formidable problems remain.

So while one cannot help being excited and enthusiastic about the discoveries and theories of the new mesons, this is a plea that the spectroscopy of the 'ordinary' mesons - although lacking a glamour quantum number - is still of great importance.

Acknowledgments

Much of this talk is based on work performed in collaboration with Roger Cashmore, Peter Litchfield and Hugh Burkardt. I am also grateful to John Ellis, Lucien Montanet and Professor Feynman for valuable discussions concerning the meson calculations.

Authors Note

The material in these lectures overlaps somewhat that covered by lectures given at Zakopane June 1975, "The Melosh Transformation: Theory and Experiment" Southampton preprint THEP 75/6-9, and an invited talk at the Argonne Summer Symposium on "New Directions in Hadron Spectroscopy", July 1975, "Lost States of the Quark Model and How to Find Them", Southampton preprint THEP 75/6-11. These lectures contain up-dated and expanded versions of the sections that overlap.

References and Footnotes


1) For more extensive discussion of the constituent quark model, together with a more complete set of references than can be given here, the reader is referred to the reviews by H.J. Lipkin (Phys. Reports C8 (1973)). J. Rosner (Physics Reports C11 (1974), and J. Weyers, Lectures at 1973 Louvain Summer School.
2) We are thus tacitly assuming 'hidden' colour for the quarks.
3) P.J. Litchfield, Review talk at the London Conference (17th Int. Conf. on High Energy Physics, London 1974, ed. J.R. Smith).
4) R.J. Hemingway, Talk at the Argonne Summer Symposium on "New Directions in Hadron Spectroscopy", July 1975; and P. Lamb private communication.
5) O.W. Greenberg, Phys. Rev. Letts 13 (1964) 598; R.H. Dalitz, Les Houches Lectures 1965 p. 253 (Gordon and Breach, New York, 1965).
6) R. Horgan, Nuclear Physics B71 (1974) 514.
7) W.D. Apel et al., Physics Letters 57B (1975) 398.
8) W. Blum et al., Physics Letters 57B (1975) 403.
9) I.J.R. Aitchison, Review Talk in "Three Particle Phase Shift Analysis and Meson Resonance Production", Proceedings of the Daresbury Study Weekend No. 8, 1975, ed. by J.B. Dainton and A.J.G. Hey.
10) U. Kruse, Review talk in Daresbury Study Weekend No. 8, 1975.
11) E.L. Berger, Review talk in Daresbury Study Weekend No. 8, 1975.
12) For cross section estimates see G.C. Fox and A.J.G. Hey, Nucl. Phys. B56 (1973) 386, and also A.C. Irving and V. Chaloupka, Nucl. Phys. B89 (1975) 345.
13) F. Wagner, contribution to Daresbury Study Weekend No. 8, 1975 - but see also D.J. Crennell's contribution.
14) See for example, G. Kane, talk at the Argonne Summer Symposium on "New Directions in Hadron Spectroscopy", July 1975.
15) See for example J.S. Bell, Schladming Lectures (1974), CERN preprint TH-1851, published in Acta Physica Austriaca; and H. Ruegg, Lectures at the XVth Cracow Summer School, Zakopane Poland (1975), to be published in Acta Physica Polonica.
16) V is usually called the Melosh Transformation, although a similar mixing operator appears in the works of many authors, in particular F. Bucella, H. Kleinert, C.A. Savoy, E. Celeghini and E. Sorace, Nuovo Cimento 69A (1970), 133 and later works.
17) J.S. Bell and A.J.G. Hey, Phys. Letters 51B (1974) 365.
18) See for example ref. 15 and other lectures at this workshop. See also H. Osborn, Nucl. Phys. B80 (1974) 90, 113; R. Carlitz et al., Phys. Rev. D11 (1975) 1234; and R. Carlitz and W.K. Tung, Chicago Preprint 1975.
19) H.J. Melosh, Phys. Rev. D9 (1974) 1095.
20) F.J. Gilman, M. Kugler, and S. Meshkov, Phys. Rev. D9 (1974) 715.
21) A.J.G. Hey, J.L. Rosner and J. Weyers, Nucl. Phys. B61 (1973) 205.
22) A.J.G. Hey and J. Weyers, Phys. Lett. 48B (1974) 69.

23) F.J. Gilman and I. Karliner, Phys. Rev. D10 (1974) 2194.

24) F.E. Close, H. Osborn and A. Thomson, Nucl. Phys. B77 (1974) 281.

25) R.J. Cashmore, A.J.G. Hey and P.J. Litchfield, Southampton Preprint THEP 74/5-6, to be published in Nuclear Physics B (1975).

26) H.J. Lipkin, Phys. Rev. D9 (1974) 1579.

27) H. Burkhardt and A.J.G. Hey, Birmingham and Southampton Preprint 1975.

28) This was emphasized by R.G. Moorhouse, Talk at London Conference 1974.

29) See for example, D. Faiman and A.W. Hendry, Phys. Rev. 173 (1968) 1720; Phys. Rev. 180 (1969) 1572, 1609; and L.A. Copley, G. Karl and E. Obryk, Phys. Rev. D4 (1971) 2844. For calculations including a more general (but still not the most general) SU(6) structure, see for example K.C. Bowler, Phys. Rev. D1 (1970) 926 and F.E. Close, L.A. Copley, and G. Karl, Oxford Preprint (1968) (unpublished).

30) A.J.G. Hey, P.J. Litchfield and R.J. Cashmore, CERN Preprint TH - 1886, to be published in Nuclear Physics B (1975).

31) D. Faiman and D.E. Plane, Nucl. Phys. B50 (1972) 379.

32) M. Jones, R. Levi Setti and T. Lasinski, Nuovo Cimento 19A (1974) 365.

33) For further discussion, see R.H. Dalitz, talk at the Argonne Summer Symposium on "New Directions in Hadron Spectroscopy", July 1975.

34) D. Faiman and J. Rosner, Phys. Lett. 45B (1973) 357; F.J. Gilman, M. Kugler and S. Meshkov, Phys. Lett. 45B (1973) 481.

35) D.E. Plane et al., Nucl. Phys. B22 (1970) 93; P.J. Litchfield et al., Nucl. Phys. B30 (1971) 125; A. Barbaro-Galtieri, LBL-1366 (1972): published in Proceedings of 1972 International Conference at Batavia (ed. J.D. Jackson and A. Roberts).

36) R.S. Longacre et al., SLAC-PUB-1390 (Rev), to be published in Physics Letters.

37) J. Prevost et al., Nucl. Phys. B64 (1974), 246.

38) R.J. Cashmore, frequently repeated private communication.

39) A. Le Yaouanc et al., Phys. Rev. D8 (1973) 2223; ibid. D11 (1975) 1272.

40) R.P. Feynman, M. Kislinger and F. Ravndal, Phys. Rev. D3 (1971) 2706.

41) This is pointed out by J. Rosner, ref. 1.

42) Several authors, recently R. Dashen and G. Kane, Phys. Rev. D11 (1975) 136, and N. Cottingham (private communication) have suggested that some candidates for radial excitations should not be classified as simple quark states. See also the talks of J.O. Dickey and R.H. Dalitz, at the Argonne Summer Symposium 1975.

43) See for example, C. Heusch and F. Ravndal, Phys. Rev. Lett. 25 (1970) 253.

44) W.J. Metcalf and R.L. Walker, Nuclear Phys. B76 (1974) 253; R.L. Walker, Phys. Rev. 182 (1969) 1729.

45) J. Babcock and J.L. Rosner, Cal Tech Preprint CALT-68-485 (1975).

46) G. Kneis, R.G. Moorhouse, H. Oberlack, A. Rittenberg and A.H. Rosenfeld, Proceedings of the 17th Int. Conf. on High Energy Physics, London 1974, ed. by J.R. Smith, contributed paper No. 957.
G. Kneis, R.G. Moorhouse, H. Oberlack and A.H. Rosenfeld, LBL-2673 (1974).

47) R.C.E. Devenish, D.H. Lyth and W. Rankin, Phys. Letters 52B (1974) 227.

48) See our Table 3 and also Table 18 of ref. 45.

49) J.L. Rosner and W.P. Petersen, Phys. Rev. D7 (1973) 747.

50) R.G. Moorhouse, Rapporteur's talk at Palermo Conf. 1975.

51) R. Carlitz and J. Weyers, Phys. Lett. 56B (1975) 154.

52) D. Faiman, Weizmann Institute preprint WIS-74/7-Ph.

53) See for example R.G. Moorhouse and N. Parsons, Phys. Lett. 47B (1973) 24; and D. Faiman, Phys. Lett. 49B (1974) 365.

54) It is crucial for P.T. Mathews and G. Feldman, I.C. Preprint 1975, who have invoked such a factor to suppress photon decays of coloured states in a colour model for the new particles. This is clearly a large and speculative extrapolation of this factor from its present domain of validity.

55) Further discussion is contained in Ref. 27. I am grateful to John Ellis of CERN and Potters Bar Institute of Technology for emphasizing this point.

56) The possibility of mixing is ignored in this "first approximation" analysis.

57) There are indications from $p\bar{p}$ annihilation into $4\pi$ via the $1^-$ channel that $\rho\varepsilon$ may be important. However, since the $\rho'(1.6)$ is well below threshold for this experiment it is difficult to make any definitive statements concerning resonance branching ratios. I am grateful to Lucien Montanet for informing of the $p\bar{p}$ analysis.

58) One can 'explain' these narrow widths for radial excitations by an othogonality argument for $H_\pi$ sandwiched between L=0 states. It is still surprising, however, that this remains valid to quite large Q values in the decay. We are grateful to Professor Feynman for some stimulating discussions concerning this point.

59) C. Bemporad, invited talk at the 1975 Int. Symposium on Lepton and Photon Interactions at High Energy, Stanford, Aug. 21-27 1975.

# CURRENT QUARKS AND CONSTITUENT-CLASSIFICATION QUARKS: SOME QUESTIONS AND IDEAS

F. E. Close

Rutherford Laboratory

## ABSTRACT

A brief introduction is given to the spin dependence of inelastic photo and electroproduction. Parton model predictions of Kuti and Weisskopf are then criticised and a paradox noted in connection with a sum rule of Bjorken. The resolution of this paradox raises several questions concerning the constituent and current quark approaches to resonance excitation. Particular attention is given to current algebra constraints, angular momentum in the nucleon, the $x \to 1$ behaviour of inelastic electroproduction, $\psi$ production and radiative decays.

## Spin dependent effects in photoabsorption - A paradox and its resolution

### 1.1 Kinematic Introduction

In inelastic scattering of polarised electrons (or muons) on polarised targets one measures a quantity $A(\nu, Q^2)$ ($\nu$, $Q^2$ the energy and squared four momentum carried by the exchanged photon) defined by

$$\frac{\frac{d^2\sigma}{d\Omega dE'}(\uparrow\Uparrow - \uparrow\Downarrow)}{\frac{d^2\sigma}{d\Omega dE'}(\uparrow\Uparrow + \uparrow\Downarrow)} = \sqrt{1-\varepsilon^2}\, A \tag{1.1}$$

where $\varepsilon \equiv (1 + 2\frac{Q^2+\nu^2}{Q^2} \tan^2 \frac{\theta}{2})^{-1}$ and longitudinal photon contributions have been neglected. The cross-sections are for lepton and target spins parallel and anti-parellel in an obvious notation, the kinematical background, if desired, can be found in Refs.1.

The physical significance of $A(\nu,Q^2)$ is more readily appreciated by considering the virtual photon-target interaction in the rest frame of the final hadronic system. The transverse photon and target nucleon can have their spins parallel (helicities anti-parallel) and hence $J_z = \pm\, ^3/_2$ or antiparallel (helicities parallel) and hence $J_z = \pm\, ^1/_2$ with the z-axis defined by their mutual direction of motion. Defining the photoabsorption cross-section in these configurations by $\sigma_{1/2}$, $\sigma_{3/2}$ (± are related by parity and so we drop the signs hereon) then

$$A \equiv \frac{\sigma_{1/2} - \sigma_{3/2}}{\sigma_{1/2} + \sigma_{3/2}} \tag{1.2}$$

1.2 Predictions for $A(\nu, Q^2)$

(i) Sign of A

In the parton model the scaling property at large $Q^2$ suggests that $A(\nu, Q^2) \rightarrow A(x)$.

From its definition it is clear that A is bounded to lie between ±1 and the first question is whether it is positive or negative or zero. Two sum rules exist which suggest that $A < 0$ for $Q^2 = 0$ but that $A > 0$ in the deep inelastic region. Namely

(a) $Q^2 = 0$: The Drell-Hearn-Gerasimov sum rule[2]

$$\frac{2\pi^2\alpha}{M^2} K^2 = \int_{\nu_0}^{\infty} \frac{d\nu}{\nu} \left[\sigma_{3/2}(\nu) - \sigma_{1/2}(\nu)\right] \tag{1.3}$$

(where K is the anomalous moment of the target) has a positive definite left hand side and hence we expect that dominantly

$$A(\nu, 0) < 0. \tag{1.4}$$

(b) $Q^2 << 0$: The Bjorken sum rule[3] for proton (P) and neutron (N) targets reads

$$\frac{1}{3}\frac{g_A}{g_V} = \int_0^1 \frac{dx}{x}\left(A^P(x)F_2^P(x) - A^N(x)F_2^N(x)\right) \qquad (1.5)$$

($F_2(x) \equiv \nu W_2(x)$ is the familiar unpolarised structure function.) If we add to this a mild assumption (namely that strange quarks can be neglected, or alternatively that diffractive effects can be neglected) then the sum rule separates into two sum rules[4-7]

$$0 = \int_0^1 \frac{dx}{x} A^N(x) F_2^N(x)$$

$$\frac{1}{3}\frac{g_A}{g_V} = \int_0^1 \frac{dx}{x} A^P(x) F_2^P(x) \qquad (1.6)$$

and so, for protons at least, we expect that

$$A(\nu, Q^2 \text{ large}) > 0 \qquad (1.7)$$

(ii) Magnitude of A

At $Q^2 = 0$ it does appear that the DHG sum rule is well saturated with the consequence that $A < 0$ is dominant[8]. For $Q^2 << 0$ there is not yet any published data in the deep inelastic region. However, preliminary statements[9] suggest that the sign of A is positive and this also seems to be the case at lower energies where a fixed

t-dispersion analysis of resonance region electroproduction suggests that A is changing sign from negative to positive at least in the second and possibly also in the third resonance regions[10].
So far the actual magnitude of A is not determined. In the quark-parton model Kuti and Weisskopf found[11]

$$A^N = 0 \quad , \quad A^P = 5/9 \tag{1.8}$$

and so $A^P > 0$ in accord with our previous guess and possibly in line with the data now appearing. Their result effectively utilises a three quark-parton structure for the nucleon and so
while they also assumed the nucleon to be a 56, L = 0 state in the classification group SU(6) ⊗ O(3). Consequently they also would have $\frac{g_A}{g_V} = \frac{5}{3}$ and so their model "satisfies" the Bjorken sum rule

$$\underset{\displaystyle 5/9}{\underset{\uparrow}{\frac{1}{3}\frac{g_A}{g_V}}} = \int_0^1 \frac{dx}{x} \underset{\displaystyle 1}{\underset{\uparrow}{F_2^P(x)}} \; \underset{\displaystyle 5/9}{\underset{\uparrow}{A^P}} \tag{1.10}$$

So things look as if they may be turning out rather well, the theoretical predictions being in the right "ballpark", though this still needs to be confirmed. Three or four years ago the few clues there were suggested that things might look quite different[12,13] and it was this possibility that led to the questions that I shall subsequently discuss. The questions are of interest whatever this data will turn out to look like. It is amusing to reflect that we might never have asked ourselves these questions had we not gotten false(?) clues from old data.

1.3 Physical Interpretation of A in the parton model: How $\sigma_{1/2}$ and $\sigma_{3/2}$ measure the distribution of quark spins $S_z$ in the target nucleon

Consider a proton $J_z = +{}^1/_2$ denoted $P\uparrow$

Then on interaction with a photon

$$\gamma(J_z = +1): \quad \gamma\uparrow + P\uparrow \Rightarrow \sigma_{3/2}$$

$$\gamma(J_z = -1): \quad \gamma\downarrow + P\uparrow \Rightarrow \sigma_{1/2} \qquad (1.11)$$

The photon flips the $S_z$ of the quarks with which it interacts. Hence, since the quarks have spin ${}^1/_2$, we find

$$\gamma\uparrow \text{ chooses } q\downarrow \longrightarrow q\uparrow$$

$$\gamma\downarrow \text{ chooses } q\uparrow \longrightarrow q\downarrow \qquad (1.12)$$

and so

$$\sigma_{3/2} \sim \gamma\uparrow p\uparrow \sim \sum_i e_i^2 q_\downarrow$$

$$\sigma_{1/2} \sim \gamma_{\downarrow} p^{\uparrow} \sim \sum_i e_i^2 q^{\uparrow} \tag{1.13}$$

hence

$$A \equiv \frac{\sigma_{1/2} - \sigma_{3/2}}{\sigma_{1/2} + \sigma_{3/2}} \simeq \frac{\sum_i e_i^2 [q^{\uparrow} - q_{\downarrow}]}{\sum_i e_i^2 [q^{\uparrow} + q_{\downarrow}]} \tag{1.14}$$

The probabilities $q^{\uparrow\downarrow}$ are for quarks with spin parallel (anti-parallel) to the target. Since naively one may expect that more quarks have their spins aligned rather than antialigned then one may expect $A > 0$.

With the probabilities that follow from an SU(6), 56 L = 0 wave function for the nucleon comprising three quarks, viz for the proton

$p^{\uparrow} = {}^5/_9 \qquad p_{\downarrow} = {}^1/_9 \qquad n^{\uparrow} = {}^1/_9 \qquad n_{\downarrow} = {}^2/_9$ ,

then one immediately finds $A^p = {}^5/_9$. Intercharging p <—>n one obtains $A^N = 0$. These are the Kuti-Weisskopf results.

1.4 Some Worries

This model also predicts that $\frac{F^N}{F^P} > \frac{2}{3}$ which is empirically not the case. In 1972 we thought[12,13] that the asymmetry might turn out negative - if so then what would be the import for the above? This brought into focus the two crucial assumptions implicit in the KW calculation. Let us examine these.

(i) The $\gamma$ interacts magnetically (i.e. flips $S_z$ but not the orbital angular momentum

This can be motivated by explicit models. It is a parton like assumption. It is not necessarily true in classification space - the most general form of interaction for a photon with single quarks may be written

$$\gamma \sim AL_+ + BS_+ + CS_zL_+ + DS_-L_+L_+ \qquad (1.14.1)$$

in an obvious notation (see Refs.[14-16] and Hey's lectures here)

(ii) <u>The nucleon is 56, L = 0, $L_z$ = 0</u>

This is well known from spectroscopy. However this is "classification-like" and not true in the parton, or current quark, representation e.g. $K \neq 0$, $\frac{g_A}{g_V} \neq \frac{5}{3}$ etc.. If one likes to associate parton models with the infinite momentum frame then on boosting the rest state 56, $L_z = 0$ to infinite momentum considerable configuration mixing arises (see e.g. Oliver's talk at this meeting.)

## 1.5 Polarisation Asymmetry - Constituent Space Calculation

Therefore Osborn, Thomson and I[14] decided to redo the KW calculation in a, hopefully, consistent fashion. Hence:

$$\begin{array}{l} (2^*)\ \text{Nucleon is } 56,\ L=0,\ L_z=0 \\ (1^*)\ \text{Photon is } AL_+ + BS_+ + CS_zL_+ + DS_-L_+L_+ \end{array} \qquad (1.15)$$

and so we are working consistently in classification space. The calculations are in appendix B (beware of misprints) of Ref.14 and reviewed in Ref.6. We found

$$F_2^N/F_2^P \geqslant 2/3$$

$$A^N = 0 \tag{1.16}$$

$$A^P = \frac{5}{9} X \; ; \; X \equiv \frac{-4AC + B^2 - D^2}{2(A^2+C^2)+B^2+D^2} \; ; \; -1 \leq X \leq 1$$

The difference with the KW results is that the photon has in effect been "rotated" from pure $S_+$ to a mixture of $S=0$ and $S=1$, the I and U spin content being untouched. Hence the spin independent result for $F^N/F^P$ is unaltered while the spin dependent $A^P$ has rotated away from $5/9$ (the quantity X being in some sense a measure of this rotation).

## 1.6 The Bjorken Sum Rule Paradox

In our calculation of section 1.5 we have

$$A^N = 0, \qquad A^P = 5/9\, X \qquad -1 \leq X \leq 1$$

but all that we can say about $\frac{dx}{x} F^P$ is that it is positive definite (working with constituent quarks we can no longer obtain the parton result relating this to the squared quark-parton charges). Hence

$$\underbrace{\frac{1}{3}\frac{g_A}{g_V}}_{\text{known positive}} = \underbrace{\int \frac{dx}{x} \{ F_2^P}_{\text{positive}} \underbrace{A^P}_{\text{could be } < 0\,?} - F_2^N \underbrace{A^N}_{0} \} \tag{1.17}$$

There appears to be nothing in the constituent space calculation that forbids $A^P < 0$. In a constituent space approach[14]

$$\frac{g_A}{g_r} \sim \langle P | \, \alpha S_z + \beta(S_+L_- - S_-L_+) | N \rangle$$

is not a function of A, B, C, D and so in some mysterious way the sum rule appears to prevent $D^2 >> B^2$. Hence the first question that will interest us:

Do current algebra sum rules place constraints on the magnitudes of A B C D, αβ a priori

and if so

Do the phenomenological results of Hey et al.[17] agree with these?

### 1.7 Polarisation Asymmetry - Current Space Calculation

Before moving on to discuss the above question let us return to section 1.5. Instead of doing the calculation in constituent space, what happens if we do the consistent calculation in current-space? A general discussion is given in Ref.6, a simple example is given here. As a candidate for the operator V which transforms constituent space wavefunctions to current space Buccella et al. suggested[18]

$$V = \prod_{i=1}^{3} V^i ; \quad V^i = 1\cos\phi + \sin\phi \left( S_+ L_- - S_- L_+ \right) \qquad (1.18)$$

The operator of Melosh[19] and of Le Youanc et al.[7,20] is also of this structure. Transforming to current space then

$$(1^{**}) \quad \text{Photon is } S_+$$

$$(2^{**}) \quad \text{Nucleon is } V \,|\, 56, L=0 \rangle \qquad (1.19)$$

and we find[6]

$$F^N/F^P = 2/3 \ ; \quad A^N = 0 \ ; \quad A^P = \frac{5}{9}\cos 2\phi \ ; \quad \frac{g_A}{g_V} = \frac{5}{3}\cos 2\phi \tag{1.20}$$

hence $A^P = \frac{1}{3}\frac{g_A}{g_V}$ and the Bjorken sum rule is satisfied. Hence our conclusion in paragraph 1.6 seems to be valid viz, current algebra sum rules can place constraints on constituent space calculations (this will be discussed in the next section). Also we see that <u>the current space wavefunction of the nucleon is more than just S wave</u>. This is qualitatively reasonable (e.g. the nucleon has $K \neq 0$ even though the partons are supposed structureless). The further discussion of this point and some consequences will be developed in sections 2.1 to 2.3.

## II. Current Algebra constraints on constituent quark calculations

### 2.1 Comparison with explicit quark models

I want to digress for a moment in order to compare the A B C D language of Hey and of this lecture with the explicit quark models mentioned by Morpurgo and others. The explicit quark models[21] have for the interaction of quark with electromagnetic field $\vec{A}$ (g,m quark g-factor and mass)

$$\sim \frac{\vec{\nabla}\cdot\vec{A}}{m_q} + \frac{g_q}{m_q}\vec{S}\cdot\vec{\nabla}\times\vec{A} + \left[\begin{array}{l}\text{spin-orbit}\\ \text{terms in ref.22}\end{array}\right] \tag{2.1}$$

and the algebraic structure clearly corresponds to that of

$$AL_+ \quad + \quad BS_+ \quad + \left[ CS_z L_+ + DS_- L_+ L_+ \right] \tag{2.2}$$

Similarly for the pion interaction one has

$$\pi \sim \alpha S_z + \beta (S_+ L_- - S_- L_+) \tag{2.3}$$

Explicit Models ~ Gamow-Teller ↑ Recoil Terms ↑

These A B C D, αβ are a priori unknown but in the explicit models can be calculated. For example

$$\alpha_{on} \sim \langle \phi_n | \exp(ikz) | \phi_o \rangle \tag{2.4}$$

where $\phi_{o,n}$ are O(3) wavefunctions for states with 0 or n excitations. In a SHO quark model one would find

$$\alpha_{on} \sim \cdots \left(\frac{k}{\alpha}\right)^n \exp(-k^2/6\alpha^2) \tag{2.5}$$

and so one could compute the various $(A \ldots D, \alpha\beta)_{ij}$. As we have seen, paradoxes can arise if these quantities are completely independent and so we ask whether sum rules can constrain or determine a priori some or all of these quantities.

## 2.2 Sum Rules and Constraints

### (a) Adler-Weissberger

Taking the commutator $[F_i^5 , F_j^5] = i f_{ijk} F_k$ between any member of the ground state 56, L = 0 and summing over any intermediate 56 or 70 then

$$\sum_{56} (\alpha^2 + \beta^2) + 2\sum_{70} (\alpha^2 + \beta^2) = 1 \tag{2.6}$$

(where $\alpha^2 \equiv \sum_{56,n} \alpha^2_{on}$ etc. diagramatically represented by

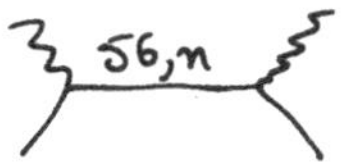


).

In the crudest approximation where only the 56, L = 0 ground state is included, then since $\beta_{oo} = 0$ we find $\alpha^2 = \alpha^2_{oo} = 1$ is required. Since

$$\frac{g_A}{g_V} = \frac{5}{3}\alpha_{oo} \tag{2.7}$$

then the old (bad) SU(6) result of $\frac{g_A}{g_V} = \frac{5}{3}$ is recovered. A non-trivial Melosh transformation requires the presence of the additional terms in the relation (2.6) and so $\frac{g_A}{g_V} < \frac{5}{3}$.

Questions that may be investigated include:-

(i) What further constraints, if any, arise from taking the commutator between 70 states?

(ii) What constraints arise from $\langle 56|[F^5, F^5]|70\rangle$ ? In this latter case I guess one will find

$$\sum_{56} (\alpha^2 + \beta^2) = 2\sum_{70} (\alpha^2 + \beta^2) \tag{2.8}$$

which is also what one finds if one demands that no-exotics are exchanged in the t-channel of the forward $\pi N \to \pi N$[14] amplitude.

(b) Cabibbo-Radicati

This may be written

$$2F_1^{\prime V}(0) = (\mu_A^V)^2 - \frac{16}{9}\mu^2 + \sum_R \frac{2}{(m_R^2 - m^2)^2}\left(\sigma^{W^+N} - \sigma^{W^+P}\right) \tag{2.9}$$

For $I = {}^1/_2$ we have $\sigma^{W^+P} \equiv 0$ which yields a positive contribution to $F_1^{\prime V}(0)$. For $I = {}^3/_2$, on the other hand, $\sigma^{W^+N} < \sigma^{W^+P}$ and a negative contribution obtains. Experimentally $F_1^{\prime V}(0) > 0$.

Inserting the resonance contributions within the SU(6) supermultiplets we have

$$2F_1^{\prime V}(0) = (\mu_A^V)^2 - \frac{16}{9}\mu^2 + \sum_{70} \frac{4}{(m_R^2 - m^2)^2}\left\{2(A^2+C^2)+B^2+D^2\right\}$$
$$+ \sum_{56} \frac{2}{(m_R^2 - m^2)^2}\left\{2(A^2+C^2)+B^2+D^2\right\} \tag{2.10}$$

and so each set of excited supermultiplets yields a positive contribution to $F_1^{\prime V}(0)$.

In explicit models A B C D may vary through the (broken) SU(6) supermultiplet. Does the $\sigma^{W^+N} > \sigma^{W^+P}$ place any quantifiable constraint upon these breaking effects?

(c) Drell-Hearn-Gerasimov Sum Rules

Take the DHG sum rule of section 1.2 and separate it into isovector and isoscalar pieces. We find

$$\left(\mu_A^V\right)^2 = \frac{16}{9}\mu^2 + \sum_{70} \frac{4}{(m_{70}^2 - m^2)^2}\left\{\frac{8}{3}AC + D^2 - B^2\right\}$$
$$+ \sum_{56} \frac{2}{(m_{56}^2 - m^2)^2}\left\{\frac{20}{3}AC + D^2 - B^2\right\} \tag{2.11}$$

and

$$\left(\mu_A^S\right)^2 = \sum_{70} \frac{4}{(m_{70}^2 - m^2)^2}\left\{\frac{1}{9}(D^2 - B^2)\right\}$$
$$+ \sum_{56} \frac{2}{(m_{56}^2 - m^2)^2}\left\{\frac{4}{3}AC + \frac{1}{9}(D^2 - B^2)\right\} \tag{2.12}$$

In the crudest approximation of retaining only the ground state supermultiplet we have

$$\left(\mu_A^V\right)^2 = \frac{16}{9}\mu^2 \quad ; \quad \left(\mu_A^S\right)^2 = 0 \tag{2.13}$$

i.e. $\mu_P = \frac{3}{2M}$ and $\mu_N = -\frac{1}{M}$ which are already pretty fair. If we add in the first excited supermultiplet 70 L = 1 (hence D=0) we find that

$$\left(\mu_A^S\right)^2 \simeq -\left(B_{01}\right)^2 = 0 \tag{2.14}$$

is the only consistent solution. Phenomenologically the result $B_{01} \simeq 0$ was noted by Gilman and Karliner[15] who raised the question whether some selection rule might be responsible. From the above sum rule we see that such a "selection rule" emerges.

Furthermore we notice that since $(\mu_A^S)^2$ is related to $[-(B)^2]$ for 56 and 70 while the positive quantities are AC and $D^2$ then if C,D=0 no consistent solution emerges. For a consistent solution, C,D $\neq$ 0 for <u>some</u> excited 56 or 70 plet.

2.3 Comments and Puzzles

In section 2.1 we saw the relation between A B C D and the explicit quark models. We commented in 2.2 that a consistent saturation of the (isoscalar) DHG sum rules required C,D $\neq$ 0. This is not a surprise since it has been known for some time that these spin-orbit like terms are required to derive the low energy theorem[(23)] which is an important part in the original derivation[(2)] of the DHG sum rule. I do not think that this has been shown before in this algebraic language[(14)].

A problem arises however when one attempts to derive the DHG sum rule in a composite model (e.g. the quark model or for a nuclear target[(24)]. In addition to the spin-orbit terms mentioned above, one finds that non-additive terms are also required[(24)]. These terms correspond to non-35 pieces in the photon or, equivalently, go beyond the single quark interaction hypothesis that motivated the algebraic structure of eq. (1.14.1) and Refs.(14-17). Hence

$$\gamma \sim AL_+ + BS_+ + CS_zL_+ + DS_-L_+L_+ + \text{other pieces.}$$

An interesting question is whether one can see this in this general language in a manner analogous to that where we found that the spin-orbit (C,D) pieces were required. Exactly where does the inconsistency arise? Perhaps a clue may be gained from Osborn's work Ref.(25).

A question for model builders - try and derive the low energy theorem or the DHG sum rule in your model. If you fail see Refs.(24 ). There are some important lessons here I believe.

2.4 Summary and discussion of questions

1. Given enough sum rules and targets can one fully constrain A B C D, $\alpha\beta$ this way?

   If the answer is yes, then I guess you need an infinite set of targets and have to solve an infinite set of equations.

2. It may be worthwhile looking at subsystems, e.g. the lowest 56 and 70 only. Can one find inconsistencies like negative equals positive?

3. Is one doing anything different from the current algebra saturation programme of previous years? Clearly there is a 1:1 correspondence. I think that this classification basis approach may yield some results which weren't obvious in previous approaches, e.g. we already saw that B(56 to 70 L=1)$\simeq$0 arises this way. This result is not obvious in previous work.

4. Instead of using the sum rules to constrain A B etc., try out your explicit models on them, e.g. SHO is tractable. Does it give any further clues?

5. Ultimately if there is a unique solution this determines the model. Should one already constrain one's model to give

   $B(56 \rightarrow 70\ 1 = 1) \simeq 0$?

   Personally I think this premature until we have a clearer understanding of the role SU(6) breaking may play, e.g.

$(\mu_A{}^S)^2 \simeq - B^2 = B_1^2 - 2B_2^2$ say for two sets of resonances 1 and 2 in the supermultiplet. In exact SU(6) $B_1^2 = B_2^2$ and the negative coefficient (and hence forced vanishing) arises. However, in broken SU(6) it may be that $B_2^2 \ll B_1^2$ and so an overall positive result would ensue. But is breaking of this sort consistent with the Cabibbo-Radicati sum rule etc.?

6. Can you get the low energy theorem and DHG sum rule in your model. Where do the non-additive terms enter? What goes wrong if you leave them out?

## III. The Nucleon in Current Quark (Parton) Space and Angular Momentum in the Nucleon

### 3.1 $\frac{g_A}{g_V}$ and its relation to angular momentum in the nucleon

In section 1.7 we noted that in a current-quark or parton basis the nucleon contains $L_z \neq 0$. We wish to discuss this further here both to show that it still obtains when one does not make specific choices for the three quark wavefunction (contrast section 1.7, in particular that choice retained the bad result $\frac{F^N}{F^P} \geqslant \frac{2}{3}$) and to show how $\frac{g_A}{g_V}$ deviating from $\frac{5}{3}$ is physically related to angular momentum or gluons in the nucleon.

We denote $q^{\uparrow\downarrow}(x)$ the probabilities to find quarks with spins up, down in a target whose spin is "up". We write $q \equiv q\uparrow + q\downarrow$, $\tilde{q} \equiv q\uparrow - q\downarrow$. Let us recapitulate the derivation of parton sum rules for unpolarised structure functions.

(i) Identities

$$F_2^P(x)/x = \frac{4}{9}p(x) + \frac{1}{9}n(x)$$
$$F_2^N(x)/x = \frac{1}{9}p(x) + \frac{4}{9}n(x) \qquad (3.1)$$

(ii) Constraints (from charges)

$$1 = \int_0^1 dx \left(\frac{2}{3}p(x) - \frac{1}{3}n(x)\right)$$
$$0 = \int_0^1 dx \left(\frac{2}{3}n(x) - \frac{1}{3}p(x)\right) \qquad (3.2)$$

yield Sum Rules for quarks

$$\int_0^1 dx\, p(x) = 2\int_0^1 dx\, n(x) = 2 \qquad (3.3)$$

(iii) Combined with identities yield sum rules for targets

$$\int_0^1 \frac{dx}{x}\left(F_2^P(x) - F_2^N(x)\right) \equiv \int dx\left(\frac{1}{3}p(x) - \frac{1}{3}n(x)\right) = \frac{1}{3} \qquad (3.4)$$

We can immediately parallel this for polarised electroproduction:-

(i) Identities

$$G_1^P(x) = \frac{4}{9}\tilde{p}(x) + \frac{1}{9}\tilde{n}(x)$$
$$G_1^N(x) = \frac{1}{9}\tilde{p}(x) + \frac{4}{9}\tilde{n}(x) \qquad (3.5)$$

(compare unpolarised)

(ii) Constraints

$$\left(\frac{g_A}{g_V}\right)^{NP} \equiv (F+D)^{Axial} = \int_0^1 dx\,(\tilde{p} - \tilde{n}) \qquad (3.6)$$

(obvious since $g_A \sim S_z$ (hence $\tilde{q}$) $*I_3$(hence $\tilde{p} - \tilde{n}$)).

(iii) Yields sum rule immediately

$$\int_0^1 dx \left(G_1^P - G_1^N\right) = \frac{1}{3}\int_0^1 dx \left(\tilde{p} - \tilde{n}\right) = \frac{1}{3}\frac{g_A}{g_V} \tag{3.7}$$

which is Bjorken's sum rule ($G_1 \sim A \times F_2$ See Ref.1).

We can go further if, neglecting strange quarks, we also include

(ii) <u>Constraint</u>

$$\left(g_A\right)^{\Xi^-\Xi^0} \equiv (F-D)^{Axial} = \int_0^1 dx\, \tilde{n} \tag{3.8}$$

yields <u>Sum Rule for quarks</u>

$$\begin{aligned} \int dx\, \tilde{p} &= 2F^{Axial} \\ \int dx\, \tilde{n} &= (F-D)^{Axial} \end{aligned} \tag{3.9}$$

and one can derive the <u>two</u> sum rules

$$\int_0^1 dx\, G_1^P(x) = \frac{1}{3}(F+D) + \frac{2}{3}\left(F - \frac{2}{3}D\right) \tag{3.10}$$

$$\int_0^1 dx\, G_1^N(x) = \qquad \frac{2}{3}\left(F - \frac{2}{3}D\right) \tag{3.11}$$

More interesting is the following result on the presence of angular momentum in the nucleon.

(1)
$$\begin{aligned} \left(\frac{g_A}{g_V}\right)^{NP} &= \int dx \left(\tilde{p} - \tilde{n}\right) \\ &\equiv (F+D)^{Axial} \\ &= \left(\frac{5}{3} + \epsilon\right) D^{Axial} \end{aligned} \tag{3.12}$$

where we defined $\left(F/D\right)^{Axial} \equiv \frac{2}{3} + \epsilon$

(2)

$$\frac{\langle S_z \rangle^{quarks}}{\langle J_z \rangle^{target}} = \int dx\,(\tilde{p} + \tilde{n})$$

$$\equiv (3F - D)^{Axial} \qquad \text{(using the sum rule e.g.3.9)}$$

$$= (1 + 3\epsilon)D^{Axial} \tag{3.13}$$

where $J_z^{target} = S_z^{quarks} + L_z^{quarks} + J_z^{gluons} = \frac{1}{2}$

Hence for $\varepsilon$ small we have finally

$$\left(\frac{g_A}{g_V}\right)^{NP} = \left(\frac{5}{3} - 4\epsilon\right)\left\{\frac{S_z^{quarks}}{1/2}\right\} \tag{3.14}$$

Then if $\varepsilon = 0$, $S_z^{quarks} = 0.37$ ($\Delta J_z = 0.13$) while experimentally $\varepsilon \simeq -0.08$ so $S_z^{quarks} = 0.30$ ($\Delta J_z = 0.20$). Note that $\frac{g_A}{g_V}$ deviation from $\frac{5}{3}$ requires either $\varepsilon > 0$ or the presence of angular momentum above that carried by the quark spins.

This result has been discussed by Sehgal[5] as well as by Close[6]. The present discussion was based on the former approach.

## 3.2 Angular momentum in the $\psi$?

The $J/\psi$ (3.1) is commonly believed to be the $^3S_1$ state of a quark-antiquark pair which we will call $c\bar{c}$ (here c refers to constituent). Is the $\psi$ (3.7) a $(^3S_1)^*$ or $^3D_1$?

The conventional argument against the latter is that $e^+e^- \not\to {}^3D_1$

since in composite models

$$\Gamma_{V\to e^+e^-} \sim |\phi(o)|^2$$
$$\sim |\vec{r}_q - \vec{r}_{\bar{q}}|^{2L} |\phi_{L=0}(o)|^2 \qquad (3.15)$$

where $\phi(o)$ refers to the spatial wavefunction of the $q\bar{q}$ at the zero separation of q and $\bar{q}$. Consequently as

$$|r_q - r_{\bar{q}}| \to 0 \text{ then } \Gamma_{V\to e^+e^-} \to 0 \text{ if } L \neq 0$$

Now, just as a proton is an S-state of constituent quarks but has an amplitude to be S, P, D ... of current quarks, so a $\psi$ which is a D-state of constituent quarks has an amplitude to be in the S state of current quark basis and hence can be produced in $e^+e^-$ annihilation.

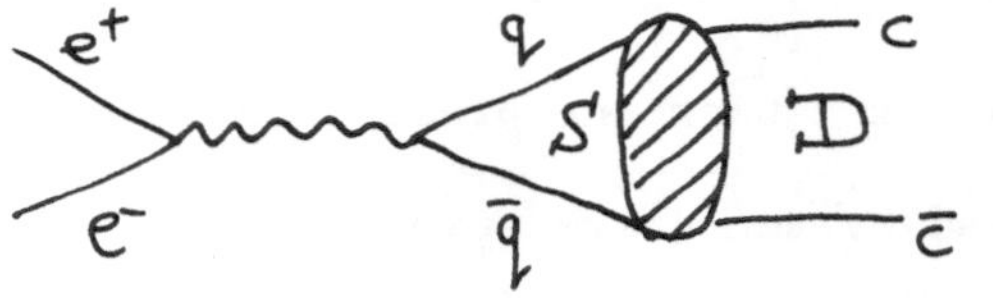


For example, relativistically the $q\bar{q}$ are four component spinors in relative S wave. The two component classification space spinors $c\bar{c}$ have S state from the large components of $q\bar{q}$ but include D state pieces in the large-small, small-small contributions. Hence the S-D mixing can be related to relativistic effects(26).

In general we therefore expect <u>both</u> $(^3S_1)^*$ and $^3D_1$ states to be produced. How can we test which (or both) is the $\psi(3.7)$? One can straightforwardly compute the helicity amplitudes or multipole amplitudes for radioactive decays like $\psi$ (3.7) $\to \gamma\chi$ (3.4) using the general structure for the photon-quark interaction described in previous sections. One finds that the contributing multipoles are[27]

| | $\chi = 0^{++}$ | $\chi = 1^{++}$ | $\chi = 2^{++}$ |
|---|---|---|---|
| $\psi(^3S_1{}^*) \to \gamma\chi$ | E1 | E1, M2 | E1, M2 |
| $\psi(^3D_1) \to \gamma\chi$ | E1 | E1, M2 | E1, M2, M3 |

hence the presence or absence of M3 contributions in $\psi(3.7) \to \gamma\chi(2^{++})$ would argue for or against the $^3D_1$ being present. The observable consequences of this and the possibility of other feasible tests are still being investigated.

## IV. The mystery of $\chi \to 1$ in deep inelastic electroproduction

### 4.1 Breaking SU(6) and configuration mixing

The SU(6) 56 assumption for the three quark wavefunction led to $F_2^N/F_2^P = \frac{2}{3}$ ; what must one do to have $F^N/F^P < \frac{2}{3}$ and what dynamical mechanism can bring this about? I don't have any fully satisfactory answer to this last question but some clue to finding the mechanism may lie in the two following ways of breaking the SU(6) to SU(3) x SU(2) or SU(3) x SU(3).

Two spin $\frac{1}{2}$ quarks can couple to S = 0 or S = 1 either of which can couple to a third quark leaving the total system with S = $\frac{1}{2}$, the S = 0(1) diquark-states being labelled by $\rho$ ($\lambda$) (cf. Dalitz in this meeting). Similar remarks apply to coupling of three I = $\frac{1}{2}$ quarks to form the I = $\frac{1}{2}$ nucleon. One then has in SU(6)

$$\underset{\sim}{56},\ I=S=\tfrac{1}{2}\,)\quad \frac{1}{\sqrt{2}}\left\{|8\rangle_\lambda|\tfrac{1}{2}\rangle_\lambda + |8\rangle_\rho|\tfrac{1}{2}\rangle_\rho\right\} \tag{4.1}$$

$$\underset{\sim}{70},\ I=S=\tfrac{1}{2}\,;\quad \frac{1}{\sqrt{2}}\left\{|8\rangle_\lambda|\tfrac{1}{2}\rangle_\lambda - |8\rangle_\rho|\tfrac{1}{2}\rangle_\rho\right\} \tag{4.2}$$

with the SU(3) x SU(2) states exhibited. In general one could envisage a mixture of these states

$$X|8\rangle_\lambda|\tfrac{1}{2}\rangle_\lambda + Y|8\rangle_\rho|\tfrac{1}{2}\rangle_\rho \tag{4.3}$$

the 56 state corresponding to X = Y = $1/\sqrt{2}$.

In this general SU(3) x SU(2) basis one finds

$$\sigma^P_{3/2} \simeq \tfrac{4}{3}X^2\,;\ \sigma^N_{3/2} \simeq 2X^2\,;\qquad \sigma^P_{1/2} \simeq 4Y^2 + \tfrac{2}{3}X^2\,;\ \sigma^N_{1/2} \simeq X^2+Y^2 \tag{4.4}$$

Hence

$$\frac{\sigma^N}{\sigma^P} = \frac{3X^2+Y^2}{2X^2+4Y^2}\ \begin{array}{l} \xrightarrow{X=Y} 2/3 \\ \xrightarrow{X=0} 1/4 \end{array} \tag{4.5}$$

and if X = Y we indeed recover the SU(6) result of $\frac{2}{3}$ while the ratio reaches a minimum of $\frac{1}{4}$ if $\chi = 0$ (ie the spin 0 diquark system alone contributes.)

For the polarisation assymetries one finds

| | | |
|---|---|---|
| $A^N = \dfrac{2 - 3\frac{\sigma^N}{\sigma^P}}{5\frac{\sigma^N}{\sigma^P}}$ | 0 | 1 |
| $A^P = \dfrac{19 - 16\frac{\sigma^N}{\sigma^P}}{15}$ | 5/9 | 1 |
| | $\frac{\sigma^N}{\sigma^P} = \frac{2}{3}$ | $\frac{\sigma^N}{\sigma^P} = \frac{1}{4}$ |

(4.6)

with the results for the two cases of interest in $\sigma^N/\sigma^P$ also shewn.

The minimum has arisen when the S = 1, $S_z = \pm 1,0$ diquark is thrown away. Can one find a dynamical mechanism for this?

An alternative way of breaking the SU(6) is to SU(3) x SU(3) where x, y are now the amplitudes to find the diquark with $S_z = 0$ or $\pm 1$. In this case

$$\sigma^P_{3/2} \simeq \frac{2}{3}x^2 \,;\quad \sigma^N_{3/2} = x^2 \,;\qquad \sigma^P_{1/2} \simeq \frac{7}{3}y^2 \,;\quad \sigma^N_{1/2} \simeq y^2 \qquad (4.7)$$

so that

$$\frac{\sigma^N}{\sigma^P} = \frac{x^2 + y^2}{\frac{2}{3}x^2 + \frac{7}{3}y^2} \quad \begin{array}{l} \xrightarrow{x=y} 2/3 \\ \xrightarrow{x=0} 3/7 \end{array} \qquad (4.8)$$

We again confirm the SU(6) result when x = y but this time the minimum value is 3/7 arising when S = 1 $S_z$ = 1 diquark is thrown away but S = 1 $S_z$ = 0 remains, in distinction to the previous example.

For the asymmetries

$$A^N = \frac{6 - 9\frac{\sigma^N}{\sigma^P}}{5\frac{\sigma^N}{\sigma^P}} \quad ; \quad A^P = \frac{27 - 28\frac{\sigma^N}{\sigma^P}}{15} \qquad (4.9)$$

one again recovers the $A^N = 0$, $A^P = 5/9$ when $\sigma^N/\sigma^P = \frac{2}{3}$ while $\sigma^N/\sigma^P \to \frac{3}{7}$ yields $A^N = 1$, $A^P = 1$ as before.

Is it possible to find a dynamical mechanism which can kill off the $S_z = 1$ diquark system?(28) This may be more meaningful under z boosts than the previous example. Although the exhibited data already appear to show $\frac{\sigma^N}{\sigma^P} < \frac{3}{7}$ one should perhaps bear in mind that nuclear physics corrections are substantial in this $x \to 1$ region and I am very cautious as to what we can really conclude about the behaviour of $\frac{\sigma^N}{\sigma^P}$ in this region.

4.2 Parton Model prediction $A^N = A^P$ as $x \to 1$

The above remarks all assumed that the $x \to 1$ behaviour was to be described by the mixture of $|8\rangle_\alpha|\frac{1}{2}\rangle_\alpha$ and $|8\rangle_\beta|\frac{1}{2}\rangle_\beta$ states, i.e. 56 and 70 $I = S = \frac{1}{2}$. These ideas have been developed by Cabibbo et al. and by Oliver et al. and are described by these authors elsewhere in this meeting. Can one make any predictions for the polarisation asymmetries without this restriction to 56, 70

$I = S = \frac{1}{2}$ states? If it is the case that $F_2^N/F_2^P \to \frac{1}{4}$ as $x \to 1$ then this is described in the quark'parton model by $\frac{n(x)}{p(x)} \to 0$.

Since $G_1^P\ (x \to 1) \sim \frac{4}{9}\tilde{p}(x)$

and $G_1^N\ (x \to 1) \sim \frac{1}{9}\ \tilde{p}(x)$

in this case (since $n\uparrow\downarrow$ , $p\uparrow\downarrow$ are probabilities the vanishing of n necessitates that of $\tilde{n}$) then $G_1^N/G_1^P \to \frac{1}{4}$.

Finally, since $\frac{G_1}{F_2} \sim A$ then

$$A^N = A^P \quad \text{if} \quad \frac{F_2^N}{F_2^P} \longrightarrow \frac{1}{4} \tag{4.10}$$

This is an unavoidable consequence of the parton model "explanation" of $\frac{\sigma^N}{\sigma^P} \to \frac{1}{4}$.

If the result $\frac{\sigma^N}{\sigma^P}\ (w \simeq 3) = \frac{2}{3}$ suggests that $A^P(w \simeq 3) = \frac{5}{9}$ and $A^N(w \simeq 3) = 0$ then if $A^N = A^P$ as $w \to 1$ some dramatic w dependence will be expected. To go beyond the prediction $A^N = A^P$ and to be able to specify the magnitude of A requires a commitment to a wavefunction or dynamical model. We gave two examples above where $A \to 1$. Examples with $A < 1$ are discussed in Ref.6.

## V. Conclusion

Clearly the spin dependence of deep inelastic scattering will teach us much about the relation between current and constituent quarks, in particular the spin rotation (Wigner rotation) aspect of this transformation. In the limit $Q^2 \rightarrow 0$ we may also test more directly the DHG as well as $B_j$ sum rules and I hope that as a result some of the questions raised here may soon be answered and, in turn, further questions be raised in their place.

## REFERENCES


1. F. E. Close, F. J. Gilman, and I. Karliner, Phys. Rev. D6, 2533 (1972).
   F. E. Close, Daresbury Report, DNPL/154.
   A.J.G. Hey, Daresbury Lecture Notes No.13.
2. S. D. Drell, A. C. Hearn, Phys. Rev. Letters, 16, 908 (1966).
   S. Gerasimov, Sov. J. Nucl. Phys. 2, 430 (1960).
3. J. D. Bjorken, Phys. Rev. 148, 1467 (1966).
4. J. Ellis and R. L. Jaffe, SLAC-PUB-1288 (1973).
5. L. Sehgal, Phys. Rev. D10, 1663 (1974).
6. F. E. Close, Nucl. Phys. B80, 269 (1974).
7. A. Le Youanc et al., Orsay Report 74/19.
8. I. Karliner, Phys. Rev. D7, 2717 (1973).
9. R. Taylor (private communication).
10. R. Devenish and D. Lyth, DESY report 1975.
11. J. Kuti and V. Weisskopf, Phys. Rev. D4, 3418 (1971).
12. F. E. Close and F. J. Gilman, Phys. Rev. Letters, 38B, 541 (1972).
13. F. E. Close, Proc. of IX Rencontre du Moriond, J. Tran. Thanh Van editor.
14. F. E. Close, H. Osborn and A. M. Thomson, Nucl. Phys. B77, 281 (1974).

15. F. J. Gilman and I. Karliner, Phys. Letters, 46B, 426 (1973).

A.J.G. Hey and J. Weyers, Phys. Letters, 48B, 69 (1974).

16. H. J. Lipkin, NAL-PUB-TH 73/62.

17. A. Hey, P. Litchfield, R. Cashmore, Nucl. Phys. (to be published).

18. F. Buccella et al., Nuovo Cimento, 69A, 133 (1970).

19. H. J. Melosh, Ph.D. thesis, Caltech.

20. A. le Youanc et al., Orsay Report 73/15.

21. D. Faiman and A. W. Hendry, Phys. Rev. 180, 1572 (1969).

L. Copley, G. Karl and E. Obryk, Nucl. Phys. B13, 303 (1969).

R. P. Feynman, M. Kislinger and F. Ravndal, Phys. Rev. D3, 2706 (1971).

22. K. C. Bowler, Phys. Rev. D1, 926 (1970).

F. E. Close, Ph.D. thesis, Oxford University.

23. F. E. Low, Phys. Rev. 96, 1428 (1954).

M. Gell-Mann and M. L. Goldberger, ibid, 1433.

24. H. Osborn, Phys. Rev. 176, 1523 (1968).

S. J. Brodsky and J. Primack, Ann. Phys. 52, 315 (1969).

F. E. Close and L. A. Copley, Nucl. Phys. B19, 477 (1970).

25. H. Osborn, Caltech 68-442.

26. Similar ideas have been put forward in a specific model by M. Krammer and H. Krasemann, DESY 75/19.

27. The results for $\psi(^3S_1{}^*)$ are implicit in F. J. Gilman and I. Karliner SLAC-PUB-1382 (1974), F. E. Close and W. N. Cottingham, CERN-TH-2009 (1975) and have also been discussed by G. Karl, S. Meshkov and J. Rosner, "Symmetries, Angular distributions in $\psi' \to \gamma\chi$ and the interpretation of the $\chi$(3400-3500) levels" (1975).

28. A recent attempt is that of G. Farrar, Caltech report (1975). See also R. Carlitz, Chicago preprint (1975) and F.E. Close, Daresbury Lecture Notes No. 12.

# SU(6)-STRONG BREAKING: STRUCTURE FUNCTIONS AND SMALL MOMENTUM TRANSFER PROPERTIES OF THE NUCLEON

A. Le Yaouanc, L. Oliver, O. Pène, and J.C. Raynal[+]

Laboratoire de Physique Théorique et Hautes Energies
Université de Paris-Sud, 91405 ORSAY CEDEX
(Laboratoire associé au Centre National de la Recherche Scientifique)

## I. The evidence for SU(6) breaking of the baryonic ground state

The SU(6) symmetry of Gürsey, Pais and Radicati[(1)] has obtained impressive achievements by classifying the low-lying baryons in the 56 representation. This led to various relations concerning the magnetic moments : $\mu_p / \mu_n = -3/2$, transition magnetic moments $\mu^*(\Delta \to p\gamma)/\mu_p = 2\sqrt{2}/3$ and axial matrix elements : $F/D = 2/3$.

Various indications of SU(6) breaking have been already encountered in the past. One very well-known is the experimental value of $|G_A/G_V| = 1.25$ against the expect 5/3. The difference is commonly understood by the chiral configuration mixing[(2)]. Others are i) the failure of $SU(6)_W$ in vertex predictions[(3)], ii) the failure of some quasi-two-body relationships[(4)] concerning the hyperon production.

In the recent years the deep inelastic electron-nucleon scattering has provided us with a most striking manifestation of symmetry breaking. The ratio of the structure functions $F_2^{en}/F_2^{ep}$, which should be $\geq 2/3$ for any x according to the 56 assignment of the nucleon, shows a monotonic decrease from 1 at $x \simeq 0$ to 1/3 at $x \simeq 1$ [(5)] (see Fig. 1).

Concerning the axial couplings and the vertex predictions, it has been shown that the observed departure from simple $SU(6)_W$ predictions could be well understood and reconciled with the 56 assignment of the nucleon :

- either in the frame of realistic quark models, by taking into account the large transverse internal momenta of quarks inside the hadrons. Two interpretations have been given, either by considering current matrix elements[6] or in a direct description of strong interaction vertices (Quark pair creation model[7], elementary meson emission models with recoil terms (ref. 10, Footnote page 4)),

- or in more formal (but practically equivalent) approaches : Melosh transformation[8], $\ell$-broken $SU(6)_W$[9].

In this case, we shall not speak any more of a breaking of SU(6) symmetry, since the simple nucleon assignment is maintained.

The $F_2^{en}/F_2^{ep}$ ratio is in our opinion quite another story. The 56 SU(6) wave function is written

$$\frac{1}{\sqrt{2}}\left(\phi'\chi' + \phi''\chi''\right) \tag{1}$$

in terms of mixed symmetry SU(3) ($\varphi$) and spin ($\chi$) wave function[10].

What is needed as shown by F. Close[11] following arguments of Feynman[12], is an enhancement of the $\varphi'$ contribution (antisymmetric with respect to two indices) for $x \to 1$.

Very interesting theoretical attempts have been made by Close et al.[13] and by Altarelli et al.[14] to interpret this enhancement. The former invoke a generalized Melosh transformation depending on $x$, while the latter invoke a (56, L = 0) + (70, L = 1) mixing suggested by the $SU(3) \otimes SU(3)$ (chiral) configuration mixing. Both have in common that the SU(6) breaking is somewhat related to the phenomenon of chiral configuration mixing.

What is our motivation for proposing another approach ?

First, we are not satisfied with these interpretations, because the connection between the observed breaking in $F_2^{en}/F_2^{ep}$ and other phenomena is not very clear. In fact Altarelli et al. do not recover the correct amount of chiral configuration mixing : the $G_A/G_V$ ratio is too small. So we feel that we have interesting, but rather ad hoc assumptions.

On the contrary, we show (next section) that there is a tight connection between deep inelastic ($2M\nu$, $-q^2 \to \infty$) and low momentum transfer or static properties of the nucleon, which has not been considered before and which extend much beyond the common SU(6) $\underline{56}$ assignment of the nucleon in both cases. This tight connection is provided by the realistic quark model (in which quarks are considered as real entities moving inside the hadron).

Using this connection, we then show (section III) that the breaking of the prediction $F_2^{en}/F_2^{ep} = 2/3$ is not truly related to chiral configuration mixings such as $((6,3)\ L_z = 0) + ((3,\bar{3})\ L_z = +1)$ as implicitly assumed in refs. 13, 14.

We propose an alternative solution, based on a true modification of the $\underline{56}$ assignment of the nucleon to a ($\underline{56}$, L = 0) + (70, L = 0) mixing, which we call SU(6) strong mixing[(15)]. We are able then to relate it to still another manifestation of SU(6) breaking, pointed out in ref. 15: the non zero charge radius of the neutron. We show that the "good" predictions of SU(6) are not much changed by this mixing (section IV). We then present a complete description of the deep inelastic scattering including gluons and pairs as in the model of Altarelli et al.[(14)] Finally our results allow us to criticize on a concrete ground the philosophy of the "current and constituent quarks" which underlies most of the recent work concerning SU(6) symmetry breaking (section VI).

We want to draw the attention to the fact that in sections III and IV we do not present the full model, but rather separate phenomena which we isolate to make their role clearer, before going into complexities.

## II. The connection between the quark model and the quark parton model

One must be aware that the idea of quarks has been applied to two rather different areas of hadron physics (not to speak about still another application which is the Gell-Mann current algebra) :

From 1964, one major development has been concerned with static properties, resonance physics and peripheral processes. (For instance, respectively the ratio $\mu_p/\mu_n = -3/2$, the decay $N^* \to N\gamma$, the density matrix of $\Delta$ in the process $\pi^- p \to \pi^0 \Delta^0$). It started with the classification of states by Dalitz[16] according to the $q\bar{q}$ and $qqq$ model and with the considerations of Morpurgo[17] on the possible non-relativistic conditions ruling the quarks inside the hadrons. An important step towards more definite predictions has been made with the introduction of the harmonic oscillator model[18], which we use in the present work as well as in all our preceding papers.

From about 1968, the experiments on the deep inelastic electron-nucleon scattering have led to a very new application of the quark idea. In the early years of the parton model, Bjorken and Paschos[19] among others already considered the eventuality that partons could be the quarks themselves. However, not only is the method of the parton model quite different from that of the ordinary quark model, but the quark interpretation was directly rejected on the ground that the number of partons is infinite. Only later, there was a growing evidence that the charged partons with spin 1/2 and charges 2/3 and -1/3 were at least compatible with experiment. This gave much support to the quark parton model of Kuti and Weisskopf[20] which described the nucleon as a 56 made up of the 3 valence quarks plus gluons and pairs.

Although for some years people were not aware of the link between such spin 1/2 fractionally charged partons and the quarks of naive quarkists, the identity between each other at least in some sense is not in serious doubt. But it was not realized that the connection may go beyond this simple idea or a common SU(6) assignment of the nucleon to a 56 multiplet of SU(6).

To establish further connection, we use the progress made in the recent years 70-75 towards a relativistic treatment of large hadron velocities and large quark internal velocities inside the hadron, in the frame of the naive quark model[6,21,22]. (* see p.171)

Our statement is that one cannot choose independently i) the structure function of the nucleon for $2M\nu, -q^2 \to \infty$ ; ii) the nucleon description at rest. They are determined (at least partially) by the same three quarks wave function via a boost operation from $P_z = 0$ to $P_z = \infty$ where the parton model is commonly understood.

The boost operation includes[21,22,23]

1°) a Lorentz contraction of the spatial wave function

$$\begin{aligned} \psi_{P_z=\infty}(\{x_i, \vec{p}_{i\perp}\}) &= N\, \psi_{P_z=0}(\{p_{iz}\sqrt{1-\beta^2}, \vec{p}_{i\perp}\}) \\ &\simeq N\, \psi_{P_z=0}(\{m_N x_i, \vec{p}_{i\perp}\}) \end{aligned} \tag{2}$$

where $\psi_{P_z=0}(p_{iz}, \vec{p}_{i\perp})$ is the usual spatial wave function and $\beta$ the bound state velocity. Note that this contraction automatically ensures, when combined with the point-like parton interaction, the scaling of the structure function.

2°) a spin boost defined by the operator acting on Dirac spinors

$$\prod_i S_i(\beta) = \prod_i \left( ch\frac{\omega}{2} + \alpha_{iz}\, sh\frac{\omega}{2} \right) \tag{3}$$

where $ch\, \omega = \dfrac{1}{\sqrt{1-\beta^2}}$ .

This last effect is trivial if the internal quark velocity is negligible : choosing Oz as the quantization axis, we get the helicity spinors at $P_z = \infty$.

With the help of this boosted wave function one can write the structure functions for the nucleon through the parton distributions $p(x)$, $n(x), \ldots, p_\uparrow(x), n_\uparrow(x), \ldots$ For instance, let $p_\uparrow(x)$ be the mean number of p-quarks with spin projection parallel to the nucleon helicity and with longitudinal fraction $x$ . In terms of the wave function at $P_z = \infty$, it can be written,

---

(*) Note that this progress has been rather disconnected from the various attempts to build formally covariant models, such as the model of Feynman, Kislinger and Ravndal (Phys. Rev. D3, 2706 (1971)), with which we differ on important points. See refs. 6, 22. In particular we keep with three-dimensional wave functions (instantaneous potential in the rest frame). The F.K.R. model does not present either the contraction, or the configuration mixing effects.

$$p_\uparrow(x) = \sum_{i=1,2,3} \int \prod_j dx_j \prod_j d\vec{p}_{j\perp}\, \delta\left(\sum_k x_k - 1\right) \delta\left(\sum_k \vec{p}_{k\perp}\right)$$
$$\times\, \psi^{\dagger}_{p\uparrow}(\{x_\ell, \vec{p}_{\ell\perp}\})\, \mathcal{P}^{(i)}_{p\uparrow}\, \delta(x - x_i)\, \psi_{p\uparrow}(\{x_\ell, \vec{p}_{\ell\perp}\}) \tag{4}$$

$\mathcal{P}^{(i)}_{p\uparrow}$ is simply the projector for the i-th quark on the spin ↑ p-quark.

This procedure is to be contrasted with that, for instance, of Altarelli et al.[14] in which the wave function at $P_z = \infty$ is chosen independently of the known quark shell model wave functions at rest.

However, when one realizes that the passage from quarks to parton models should be, according to (4), of purely kinematical nature, two questions are raised :

A. It would lead to a description of the nucleon by only three valence quarks. How is this compatible with the evidence for $q\bar{q}$ and gluons contributions ?

A possible answer can be found in the work of Altarelli et al.[14], where it is suggested that the three quarks wave function is a wave function for dressed quarks, which are themselves composed of bare quarks and $q\bar{q}$ and gluons. These are the true partons. Then the parton distributions are written as convolution products, such as

$$p(x) = \int_x^1 \left\{ p_0(z)\left[\phi^a\left(\frac{x}{z}\right) + \phi^c\left(\frac{x}{z}\right)\right] + n_0(z)\, \phi^c\left(\frac{x}{z}\right)\right\} dz \tag{5}$$

where $p_0(z)$ are the former $p(x)$ of formula (4) and $\varphi^a$ and $\varphi^c$ are the valence and core contributions to the dressed quark.

This is different from the standard quark parton picture, where the $q\bar{q}$ and gluons are directly added to three valence quarks. The advantage is that, in this new picture, we have three entities keeping their individuality inside the nucleon. This picture is in agreement with what we have found in the resonant electroproduction[22]. The quarks which constitute the nucleon must themselves have some structure, described in that case by form factors coming from Vector Meson Dominance.

The complication induced by (5) is needed to get correct absolute magnitudes for the scaling functions. With only three quark partons, one could not get the correct position and height of the quasi-elastic peak; to get it, we need neutral gluons. Moreover, to reproduce the low $x$ behavior, we need a $q\bar{q}$ contribution. However, as far as ratios of structure functions such $F_2^{en}/F_2^{ep}$ are concerned, and for large $x$, the three quark parton model defined by (4) is already sufficient as we shall see later (see Figure 1). Then, as we want to discuss the various effects involved with the maximum of clarity, we first discuss the problem of SU(6) breaking in the frame of relation (4) (sections III, IV).

B. One may wonder, more generally, how the knowledge of $s, -t \simeq 0$ phenomena (nucleon wave function at rest) could imply the knowledge of the deep inelastic region. Does it mean that nothing is new in this latter region ?

The answer is that parts of the wave function which are unimportant at $t \simeq 0$ may play a major role at $2M\nu$, $-q^2 \to \infty$. For instance, we show in Section IV that a small mixing of the nucleon wave function ($\underline{56}$, L = 0) + ($\underline{70}$, L = 0), can explain the behavior of $F_2^{en}/F_2^{ep}$ for $x \to 1$, while not modifying much the SU(6) predictions for static properties of the nucleon. Furthermore, the deep inelastic scattering makes it possible to test the quark structure itself in a more detailed manner than low $t$ scattering. On the other hand, it is also possible to go from the knowledge of deep inelastic scattering to various consequences concerning the low $t$ scattering which may be tested by more refined experiments ; for instance from the ($\underline{56}$, L = 0) + ($\underline{70}$, L = 0) mixing of Sec. IV, introduced to explain the ratio $F_2^{en}/F_2^{ep}$, one infers the presence of a charge radius for the neutron which is indeed observed although it is a rather small phenomenon (since the form factor $G_E^n(q^2)$ is, in the average, very small) and in some sense "unimportant".

We now pass to the consequences of relations of the type (4) with various hypotheses.

## III. Nucleon in a pure 56 at rest, with relativistic quark motion : $F_2^{en}/F_2^{ep}$ stays at 2/3

A. Let us look more carefully at the spin part of the 56 nucleon wave function

$$\psi_{\vec{P}=0} = \frac{1}{\sqrt{2}}\left(\phi'\chi' + \phi''\chi''\right)\psi \tag{6}$$

$\psi$ being the spatial part.

One gets the usual static results such as $\mu_p/\mu_n = -3/2$ and $|G_A/G_V| = 5/3$ when one describes the quark spin by Pauli spinors and one takes as the magnetic moment operator $\vec{\mu} \simeq \sum_i e_i \vec{\sigma}_i$ and as the axial operator $\sum_i \tau_i^{\pm} \sigma_{iz}$.

However, as we have repeatedly emphasized[6], it is now clear that with light effective quarks, as are necessary to describe the effective interactions of quarks (we do not state anything about the bare mass of the quark), the quark velocity should not be small, but rather close to unity. In fact, it may be estimated from the mass spectrum interval $\Delta E$ and the wave function radius $R$ to be :

$$v \simeq R\Delta E \simeq 1 .$$

Then, the quark spin should be described relativistically by Dirac spinors $u_i$, and the above nonrelativistic calculations of the magnetic moment and the axial charge should be replaced by the calculation $\sum_i e_i \bar{u}_i \gamma_i^{\mu} u_i$ and $\sum_i \bar{u}_i \tau_i^{(+)} \gamma_i^{\mu} \gamma_i^5 u_i$ sandwiched between the nucleon wave functions. This is at least the simplest prescription maintaining the independent particle model (See ref. 24).

Equivalently, one can say that, to calculate a matrix element, one takes the operator in terms of Dirac $\gamma_i^{\mu}$'s and one uses the same wave functions as before (6) but with the replacement in the spin part :

$$\chi_i \text{ (Pauli spinors)} \rightarrow u_i \tag{7}$$

What comes out is a correction to the various matrix elements ; one striking example is :

$$|G_A/G_V| = \frac{5}{3}(1 - 2\delta) \tag{8}$$

The correction is characterized by $\delta$, which measures the squared magnitude of the small components in the Dirac spinors.

For definiteness, we take

$$u_i \sim \begin{pmatrix} \chi_i \\ \mu_q \, \vec{\sigma}_i \cdot \vec{p}_i \, \chi_i \end{pmatrix} \tag{9}$$

where $\mu_q$ is some parameter (of the order $1/2m_q$). This is suggested by the case of free quarks. We have then $\delta \sim \mu_q^2/R^2$ which is indeed large and leads to $|G_A/G_V| \simeq 1$. (We do not want to discuss the value more quantitatively in view of the uncertainties).

For a more formal presentation of the same wave functions, see Kellett[(25)]. The wave functions are however also introduced by hand.

B. We now make the boost prescription (3) :

The quark spinor $u_i$ goes to

$$\frac{1}{\sqrt{2}}\left[1 + \mu_q m_N \left(\alpha_i - \frac{1}{3}\right) + \mu_q W_i^+ p_i^- + \mu_q W_i^- p_i^+\right] \begin{pmatrix} \chi_i \\ \sigma_{iz} \chi_i \end{pmatrix} \tag{10}$$

where $W_i^{\pm} = \frac{\sigma_{ix} \pm i\sigma_{iy}}{2}$ , $p_i^{\pm} = \mp \frac{1}{\sqrt{2}}(p_{ix} \pm i p_{iy})$ .

If the internal velocities were negligible, we would have only :

$$\frac{1}{\sqrt{2}} \begin{pmatrix} \chi_i \\ \sigma_{iz} \chi_i \end{pmatrix}$$

that is, the helicity spinor with the same spin projection as at $P_z = 0$. Then the boosted wave function would be exactly the same as at rest, in terms of helicity spinors.

Considering the classification group SU(3) ⊗ SU(3), which is generated by $\lambda_i^\alpha$, $\lambda_i^\alpha \sigma_{iz}$ ($\alpha = 1,\ldots,8$) at $P_z = \infty$, we get, from a $\underline{56}$ of SU(6), the classification under SU(3) ⊗ SU(3) simply by restricting to the subgroup of spin with $\sigma_{iz}$ instead of $\sigma_{iz}$, $\sigma_{ix}$, $\sigma_{iy}$ for the full group.

We then obtain a (6,3) representation for an octet and a decuplet with helicity + 1/2.

This is no more true when one considers the other terms in the formula (10). The $W_i^\pm$ operators completely change the SU(3) ⊗ SU(3) structure since they reverse the helicity. The new representations are connected with momentum factors coming from the $p_i^\pm$, such as to maintain $J_z = + 1/2$.

One gets [15,6], restricting to the octet,

$$\begin{aligned}\psi_\uparrow(\{x_i, \vec{p}_{i\perp}\}) = {} & A(L_z=0)\,|(6,3)\rangle_8 + A'(L_z=+1)\,|(3,6)_8\rangle \\ & + B(L_z=+1)\,|(3,\bar{3})_8\rangle + B'(L_z=0)\,|(\bar{3},3)_8\rangle \\ & + \Gamma(L_z=-1)\,|(8,1)\rangle + \Gamma'(L_z=+2)\,|(1,8)\rangle\end{aligned} \tag{11}$$

where the weight $\alpha^2$, $\alpha'^2$, $\beta^2$, $\beta'^2$, $\gamma^2$, $\gamma'^2$ of the various representations can be calculated with the harmonic oscillator wave functions [For details, see our paper[6]]. $\alpha^2$ is the norm of the $(6,3)_8$ component,... and $\gamma'^2$ is the norm of the (1,8). The expansion (11) in terms of SU(3) ⊗ SU(3) at $P_z = \infty$ is of course very suitable for analyzing the properties of the hadronic currents, since one irreducible representation of SU(3) ⊗ SU(3) has simple current matrix elements.

For axial couplings, one gets in particular for the $1/2^+$ octet(*)

$$F = \frac{2}{3}(\alpha^2 - \alpha'^2) + (\gamma^2 - \gamma'^2) \; ; \; D = (\alpha^2 - \alpha'^2) + (\beta^2 - \beta'^2)$$

$$\left\{\begin{aligned} |G_A/G_V| &= \frac{5}{3}(\alpha^2 - \alpha'^2) + (\beta^2 - \beta'^2) + (\gamma^2 - \gamma'^2) \\ F/D &= \frac{2}{3}\end{aligned}\right. \tag{12}$$

(*) We assume that $(g_A)_q = 1$, which is not theoretically necessary. See chapter VI.

The fact that F/D remains at its naive SU(6) value derives from the fact that, as shown in (8), the SU(6) naive magnitudes for the axial charges are simply renormalized by a common factor $1-2\delta$. It implies $(\gamma^2 - \gamma'^2) = \frac{2}{3}(\beta^2 - \beta'^2)$.

The present explanation of the chiral configuration mixing for the nucleon is fundamentally the same as that proposed by Melosh with rather far-fetched arguments, and which is often referred to as the distinction between current and constituent quarks. From our point of view, the phenomenon under question does not raise serious conceptual difficulties, as it is simply the philosophy of the Dirac equation. Now, we think that this is not the whole matter, and that other phenomena are involved in the configuration mixing (see sections IV, V). In section VI, we shall return to the distinction between current and constituent quarks. It must be kept in mind, however, that the phenomenon of relativistic velocities and small components is responsible for a large part of the configuration mixing.

The $SU(6)_W$ "breaking" manifested in particular in the axial charges is the base for the SU(6) mixing at $P_z = \infty$ assumed by Altarelli et al.

$$(8, \frac{1}{2}^+) \in (\underline{56}, L=0) + (\underline{70}, L=1)$$

However, the boost operation does not lead, according to us, to a definite $SU(6) \otimes O(3)$ mixing scheme at $P_z = \infty$, but only to an $SU(3) \otimes SU(3)$ mixing scheme, with $L_z \neq 0$ components. Although (11) has common features with a $(\underline{56}, L=0) + (\underline{70}, L=1)$, it is not equivalent (see our paper[6]) (and this is why Altarelli et al. can get $F_2^{en}/F_2^{ep} \neq 2/3$). In fact (11) does not mean a breaking of SU(6), but rather a failure of a too naive use of SU(6).

C. We next consider the consequences of (10,11) for the structure functions.

The effect of the small Dirac components will essentially influence the polarized structure functions, and not the unpolarized ones, since it is mainly a spin effect.

In fact, when one normalizes the wave function properly, one finds exactly the same unpolarized structure function as if $\mu_q \simeq 0$.

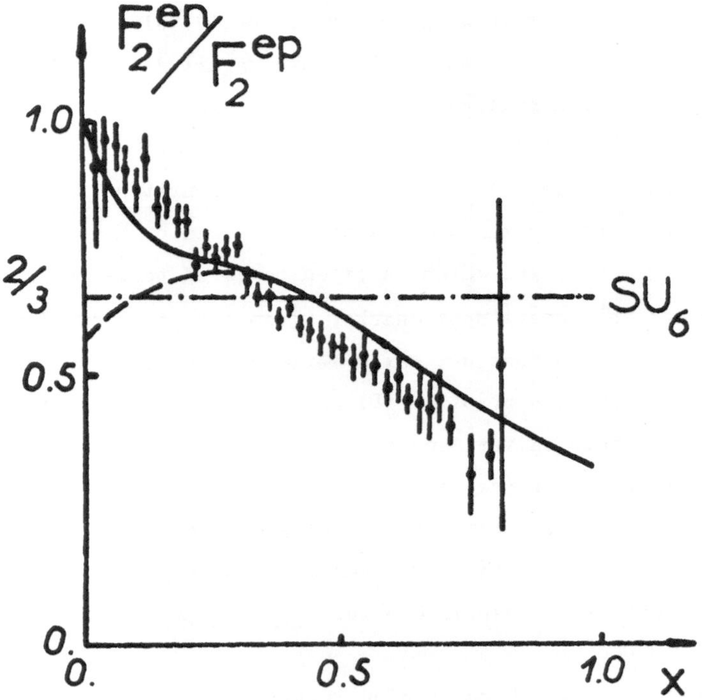


Fig. 1. The ratio $F_2^{en}/F_2^{ep}$ in the various approximations : three quarks with SU(6) unbroken (dotted-dashed line) ; valence contribution with SU(6) broken (dashed line) ; valence + sea contributions (full line). Date are taken from ref. 5.

$$F_2^{ep} = xN \exp\left[-\frac{3R^2 m_N^2}{2}\left(x-\frac{1}{3}\right)^2\right]$$

where N is a normalization factor and $R^2$ is the harmonic oscillator model radius (for comparison with experiment see Fig. 2). Also we get $F_2^{en}/F_2^{ep} = 2/3$. One sees that although a SU(3) ⊗ SU(3) representation mixing is generated at $P_z = \infty$ according to (11), by the effect of the Dirac small components, it does not lead to anything like the desired $F_2^{en}/F_2^{ep}$ behavior at $x \to 1$. This has exactly the same origin as the fact that F/D remains at its naive SU(6) value 2/3. The relative weight of $\varphi'$ and $\varphi''$ in (6) is unaltered under the substitution (7,9).

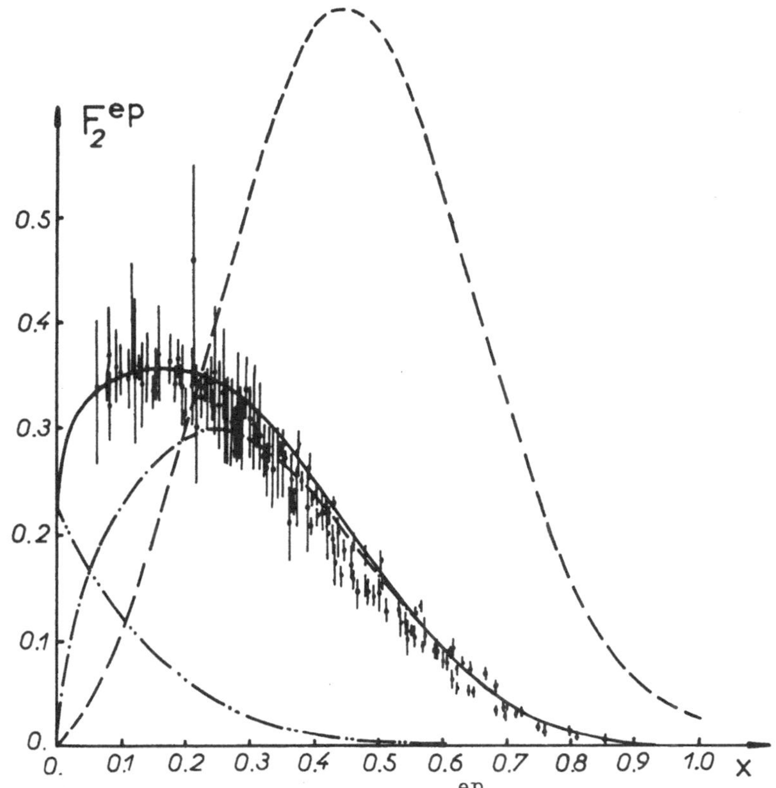


Fig. 2. The proton structure function $F_2^{ep}$. Data are taken from ref. 33. The dashed line is the wave function prediction in the approximation of three quarks. The dotted-dashed line is the valence contribution once neutral gluons are introduced. The dotted-dotted-dashed line is the $q\bar{q}$ sea contribution. The sum of both is the full line.

Then, inasmuch as we keep with the original Melosh transformation, we conclude that the behavior of $F_2^{en}/F_2^{ep}$, at $x \to 1$, has nothing to do with Melosh. This has been also emphasized clearly by S. Kitakado[26]. This is the reason why we disagree with the philosophy of Close[13], although we agree with much of its practical conclusions. His $x$-dependent transformation between current and constituent quarks is in fact disconnected from the Melosh original ideas.

For polarized structure functions, following the calculation of Kuti and Weisskopf, we find nontrivial consequences of (11). Defining the scaling limit

$$\frac{1}{2\pi}\left[\nu\, d(q^2,\nu)\right]_{p,n} \longrightarrow A^{p,n}(x)\, G(x)$$

for the $d(q^2,\nu)$ structure function, one gets

$$\begin{cases} A^p = \frac{5}{9}(\alpha^2 - \alpha'^2) - \frac{1}{9}(\beta^2 - \beta'^2) + (\gamma^2 - \gamma'^2) \\ A^n = -\frac{4}{9}(\beta^2 - \beta'^2) + \frac{2}{3}(\gamma^2 - \gamma'^2) \end{cases} \tag{13}$$

or

$$\begin{cases} A^p = \frac{1}{3}(F+D) - \frac{2}{3}\left(\frac{2}{3}D - F\right) = \frac{1}{3}\left|\frac{G_A}{G_V}\right| < \frac{5}{9} \\ A^n = -\frac{2}{3}\left(\frac{2}{3}D - F\right) = 0 \end{cases} \tag{14}$$

with the help of (12).

Instead of $A^p = 5/9$, $A^n = 0$, according to Kuti and Weisskopf.

Then $A^p - A^n$ satisfies the Bjorken sum rule[27]. The relations (14) have been found independently by J. Ellis and R. Jaffe[28]. $A^p = \frac{1}{3}|G_A/G_V|$, $A^n = 0$ is the result of Close[13]. Finally, in our scheme, one gets automatically the results of Sehgal[29], concerning the "orbital momentum composition" of the nucleon

$$S_z = \frac{1}{2}(3F - D)$$

$$L_z = \frac{1}{2}(1 - 3F + D).$$

However, the quantities $S_z$ and $L_z$ at $P_z = \infty$ <u>are not</u> the $\vec{S}_q$ and $\vec{L}_q$ of the realistic quark model defined at $P_z = 0$. $\vec{L}_q = 0$ for the three quarks inside the nucleon, even with a Dirac equation (see P.N.Bogoliubov ref. 24).

D. Finally let us quote the following unsolved problem.

The magnetic moment of the proton comes out

$$\mu_p = \frac{2}{3}\mu_q\left(1-\frac{2}{3}\delta\right) + \frac{1}{2m_N}\left(1-2\delta\right) + \kappa_q\left(1-2\delta\right)$$

To explain the absolute magnitude of $\mu_p$, one needs a large anomalous quark magnetic moment $\kappa_q \simeq 1.3/2m_N$ (see also Kellett[25]). How can this be reconciled with the fact that the partons are point like and have a minimal electromagnetic interaction ?

The problem is not modified by the introduction of a small SU(6) mixing in Section IV or of an unpolarized parton sea in Section V.

## IV. SU(6) mixing ($\underline{56}$, L = 0) + ($\underline{70}$, L = 0) for the nucleon, without the small components effect. Description of the $F_2^{en}/F_2^{ep}$ behavior. The neutron charge radius.

A. What we need really for explaining the behavior of $F_2^{en}/F_2^{ep}$ is a change in the relative amount of $\varphi'$ and $\varphi''$. This can only be due to a mixing of representations describing the nucleon. Let us now discuss the possible mixing schemes, without taking into account the complication of Dirac spinors which seems unessential. The discussion is also simpler at $\vec{P} = 0$.

We understand the mixing as a mixing between the states of the harmonic oscillator model[18]. Parity leads to a mixing between the states $n = 0$ and $n = 2$. What are the $n = 2$ $P_{11}$ states ? There are four multiplets containing an octet with spin $\frac{1}{2}^+$ :

$$(\underline{56},\ L = 0^+),\quad S_q = 1/2 \tag{15}$$

$$(\underline{70},\ L = 0^+),\quad S_q = 1/2 \tag{16}$$

$$(\underline{20},\ L = 1^+),\quad S_q = 1/2 \tag{17}$$

$$(\underline{70},\ L = 2^+),\quad S_q = 3/2 \tag{18}$$

Mixing with (15) does not break SU(6). Mixing with (17) or (18) will not give a sizable effect, because it will not give an interference term between the two wave functions. The desired effect must come from (16).

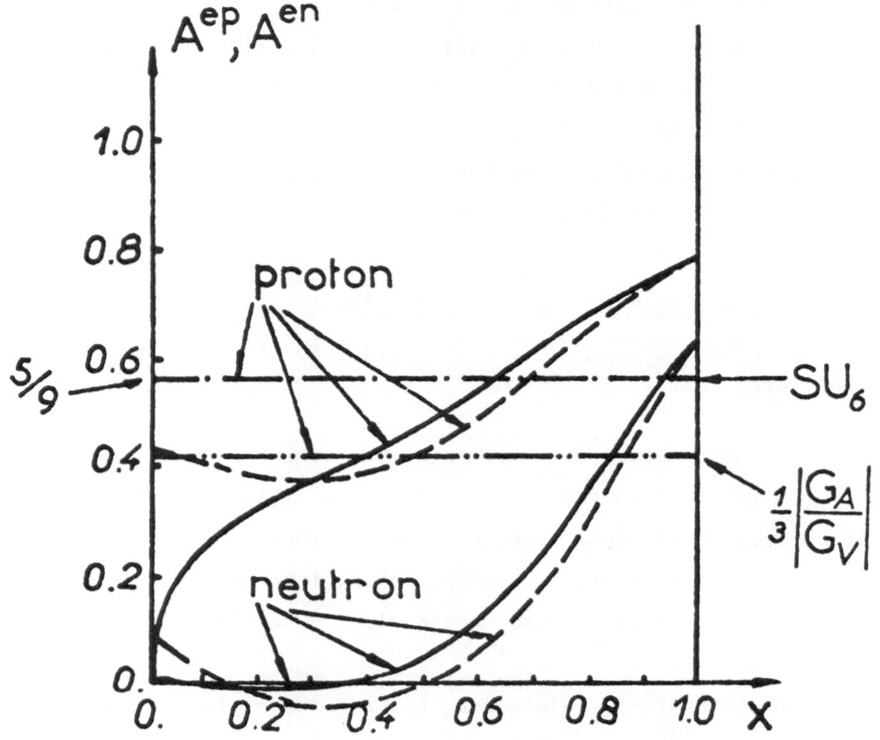


Fig. 3. Prediction for the quantity $A(x) = x\, G_1(x) / F_2(x)$ (asymmetry) for proton and neutron targets in the various approximations: three quarks with SU(6) static unbroken (dotted-dashed line for the proton, null value for the neutron) ; three quarks with SU(6) exact but the spin part written in terms of Dirac spinors (dotted-dotted-dashed line for the proton, null value f the neutron) ; valence contribution with SU(6) broken (dashed lines) ; valence + sea contributions (full lines).

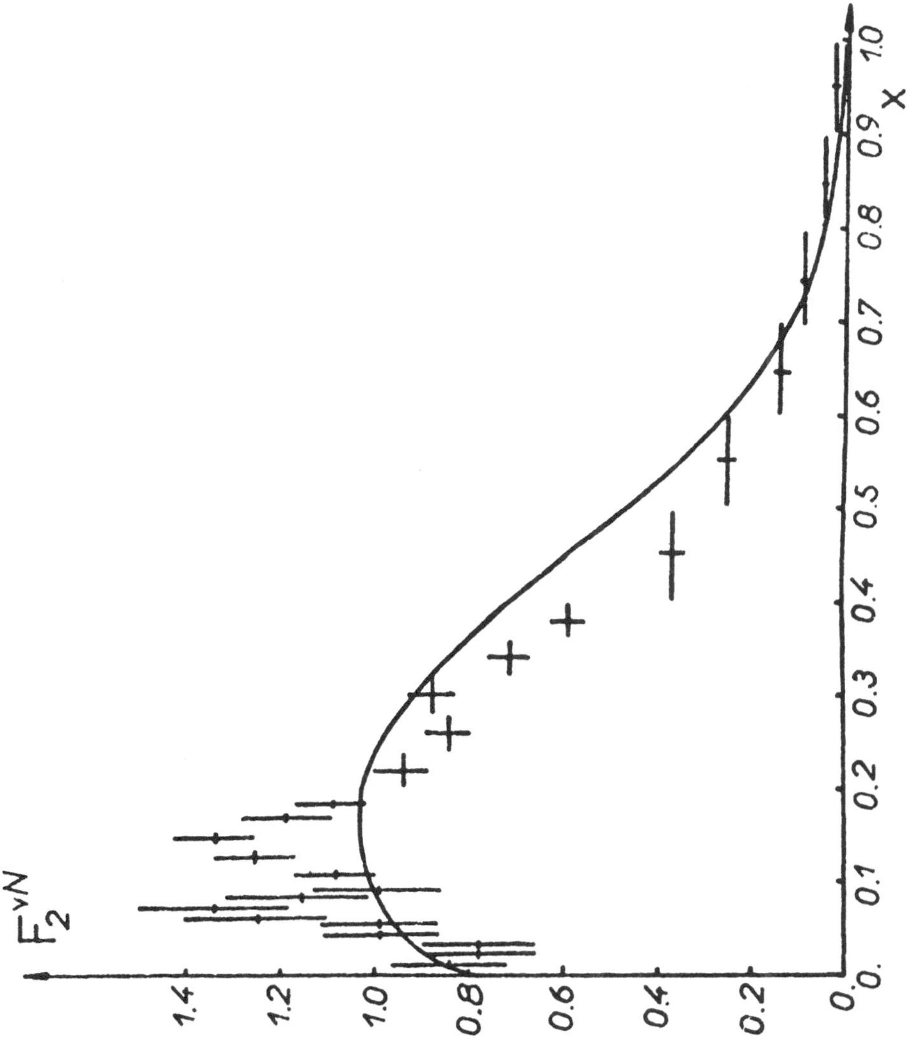


Fig. 4. The prediction of our model for the average $F_2^{\nu N} = \frac{1}{2}(F_2^{\nu p} + F_2^{\nu n})$ compared to the data of the Gargamelle Neutrino Collaboration (ref. 34).

For the moment, we have no dynamical understanding of this mixing scheme. So we just state that, for some reason, the nucleon is described by

$$\psi = \cos\varphi\, |\underline{56}, 0^+, S_q = 1/2\rangle_{n=0} + \sin\varphi\, |\underline{70}, 0^+, S_q = 1/2\rangle_{n=2} \tag{19}$$

and we test this by its practical consequences on the properties of the nucleon.

Let us just write the two components, restricted to the octet, in an explicit form :

$$|56, 0^+, S_q = 1/2\rangle_{n=0} = \frac{1}{\sqrt{2}}(\chi'\phi' + \chi''\phi'')\,\psi^S$$

$$|70, 0^+, S_q = 1/2\rangle_{n=2} = \frac{1}{2}\left[(\psi''\chi' + \psi'\chi'')\,\phi' + (\psi'\chi' - \psi''\chi'')\,\phi''\right]$$

where the spatial wave functions are :

$$\psi^S = N_0 \exp\left[-\frac{R^2}{2}(\vec{p}_\rho^{\,2} + \vec{p}_\lambda^{\,2})\right]$$

$$\psi'' = N_1 (\vec{p}_\rho^{\,2} - \vec{p}_\lambda^{\,2}) \exp\left[-\frac{R^2}{2}(\vec{p}_\rho^{\,2} + \vec{p}_\lambda^{\,2})\right]$$

$$\psi' = N_1 (2\vec{p}_\rho \cdot \vec{p}_\lambda) \exp\left[-\frac{R^2}{2}(\vec{p}_\rho^{\,2} + \vec{p}_\lambda^{\,2})\right]$$

$$\vec{p}_\rho = \frac{1}{\sqrt{2}}(\vec{p}_1 - \vec{p}_2)\,, \quad \vec{p}_\lambda = \frac{1}{\sqrt{6}}(\vec{p}_1 + \vec{p}_2 - 2\vec{p}_3)$$

B. Now we look at the ratio $R^{np}(x) = F_2^{en}/F_2^{ep}$

$$R^{np} = \frac{\langle \phi_0'|\,\hat{e}_{(3)}^2\,|\phi_0'\rangle\, C_A(x) + \langle \phi_0''|\,\hat{e}_{(3)}^2\,|\phi_0''\rangle\, C_S(x)}{\langle \phi_+'|\,\hat{e}_{(3)}^2\,|\phi_+'\rangle\, C_A(x) + \langle \phi_+''|\,\hat{e}_{(3)}^2\,|\phi_+''\rangle\, C_S(x)} \tag{20}$$

If $\varphi = 0$, $C_A = C_S$ and $R^{np} = 2/3$. But if $\varphi \neq 0$, $C_A \neq C_S$ and they have a different $x$ dependence. So $R^{np}$ varies with $x$.

The interference term $\sin\varphi \cos\varphi \langle \psi'' | \delta(x - x_3) | \psi^S \rangle$ contributes with opposite sign to $C_A$ and $C_S$, so that there can be a cancellation in the numerator of (20).

For $\varphi \simeq -20°$, one has the largest possible cancellation and $R^{np} \to 1/3$ when $x \to 1$.

This is compatible with the data, and, as one sees in Fig. 1, one is then able to explain $R^{np}(x)$ for $x \geq 0.25$ (for small $x$, one must resort to $q\bar{q}$ pairs).

In Fig. 3, one sees the asymmetries $A^{en}$, $A^{ep}$, which are also much variable over the range $1 \geq x \geq 0.25$, and get very close to each other for $x \to 1$ (30). The predictions of the naive SU(6) quark parton model or of Section III are completely removed.

We conclude that the relatively small mixing (19) is able to describe completely the behavior of the ratios of structure functions of the nucleon (for large $x$).

The reason for this relies

1°) on the existence of an interference term between the two wave functions, which is linear in $\mathrm{tg}\,\varphi$,

2°) on the rapid growth of this term due to the polynomials which describe the radial excitation.

C. On the contrary, it is easy to show that the static properties of the nucleon are only slightly modified. The most usual quantities have corrections quadratic in $\mathrm{tg}\,\varphi$, which is small. One has:

$$\mu_p \to \mu_p \left(\cos^2\varphi + \frac{1}{3}\sin^2\varphi\right) \tag{21}$$

$$\mu_p/\mu_n = -\frac{3}{2}\left(1 + \frac{1}{3}\,\mathrm{tg}^2\varphi\right) \tag{22}$$

$$|G_A/G_V| = \frac{5}{3}\left(1 - \frac{4}{5}\sin^2\varphi\right) \tag{23}$$

$$F/D = \frac{2}{3}\left(1 + \frac{1}{2}\,\mathrm{tg}^2\varphi\right) \tag{24}$$

So, we see that the mixing (19) is able to explain the large breaking of SU(6) at $x \to 1$ while retaining the classical SU(6) results for the static properties. It is exactly the desired effect.

Note that, from (23), the mixing responsible for the behavior of $R^{np}(x)$ is not able to explain the deviation of $|G_A/G_V|$ from 5/3. The same conclusion has been obtained by Altarelli et al. (14), with the mixing ($\underline{56}$, $L=0$) + ($\underline{70}$, $L=1$) at $P_z = \infty$.

The deviation of $|G_A/G_V|$ is not mainly related to the deviation of $R^{np}(x)$ from 2/3. It is mainly an effect of the small components (Section III). And so is the chiral configuration mixing.

Note also that the prediction $\mu_p/\mu_n = -3/2$, $F/D = 2/3$ are corrected in a sense opposite to the experimental numbers, but since the correction is small, the predictions are as good as before. The correction is simply irrelevant.

## V. Introduction of the gluons and of the $q\bar{q}$ sea
## Final results combining the various effects

In sections III and IV, we have made precise the structure of the three quark wave function describing the nucleon in terms of, so to say, dressed quarks. According to (5), there remains to determine the structure function of the quarks $\varphi^a$ and $\varphi^c$.

In principle one should be able to relate these structure functions to what we know of the quark structure in the small momentum transfer region, in just the same way as we have related the wave function of the nucleon to the wave function used in the resonance and in the peripheral region. An example of such a connection is given by the link between the Regge behavior of quark interactions and the behavior of the quark structure function at $x \to 0$ in the work of Altarelli et al. (which formulate at the quark level earlier suggestions of Kuti and Weisskopf). For the present time, we have not been able to do better, and we adopt almost entirely the description of the quark structure functions given by Altarelli et al.

Before presenting it, let us note however the strong indications for the presence of a quark structure in the low t region, which have in general been underestimated by the workers in the field of quark models.

We do not want to insist on the eventual presence of a $(g_A/g_V)_q$ ratio $\neq 1$ or of an anomalous magnetic moment $\varkappa_q$, since these are much controversial questions, and moreover, we do not propose any coherent answer. Much more unambiguous in our mind is the argument taken from the electromagnetic form factors of the nucleon.[21,22] Only half of the slope $d\,G_E^p / dq^2$ is explained by the $t$ dependence of the transition form factor $< \psi_N | e^{i q r_3} | \psi_N >$, with $R^2 = 6\text{-}8\ \mathrm{GeV}^{-2}$ as deduced from various phenomena[34], including the present work.

To explain the value $< r_p^2 > \approx 17\ \mathrm{GeV}^{-2}$, one must include a quark form factor contribution, which could be given by the VMD pole term $1/(t - m_\rho^2)$. The same argument may be applied to the two-body strong scattering : only half of the $t$ dependence may be attributed to the transition form factors, the remaining part must come from the quark-quark Regge scattering. If there are such important structure effects of the quark for small $t$, no wonder that it also plays a very important role in the deep inelastic region.

For the structure functions $\varphi^a$ and $\varphi^c$, Altarelli et al. propose the following expressions

$$\phi^a(x) = \frac{\Gamma(A+\frac{1}{2})}{\Gamma(A)\,\Gamma(\frac{1}{2})} \cdot \frac{(1-x)^{A-1}}{\sqrt{x}} \tag{25}$$

$$\phi^c(x) = \frac{C}{x}\,(1-x)^{D-1} \tag{26}$$

The behavior for $x \to 0$ is suggested by Regge arguments. The gluon and $q\bar{q}$ sea (being responsible for the diffraction) must behave like $A\,\frac{dx}{x}$. The valence contribution $\varphi^a(x)$ must behave like ordinary Regge exchange, i.e., $1/\sqrt{x}$. From the behavior of the sea at $x \to 0$, one deduces also the behavior of $\varphi^a(x)$ at $x \to 1$, $(1-x)^{A-1}$. Finally $D$ is an ad hoc parameter such that $D \gg A$.

The parameters are estimated (and not fitted), by the required agreement with the deep inelastic data.

For the final results, one includes :

(i) The effect of the small components,

(ii) ($\underline{56}$, $L=0$) + ($\underline{70}$, $L=0$) mixing at rest,

(iii) the effect of the quark structure.

$R^2$ (i.e. the nucleon wave function radius) is fixed to $8\ \mathrm{GeV}^{-2}$, (and the results are rather sensitive to it), by the behavior of $F_2^{ep}(x)$. This is compatible with other determinations[32].

$\varphi$ (the mixing angle) is fixed by the minimization of $R^{np}(1)$ ($\varphi \simeq -20°$) $\mu_q^2/R^2$ (magnitude of the small components) is then fixed by the value of $G_A/G_V$ ($g_A = 1$) : $\mu_q^2/R^2 = 0.10$.

From inspection of the curves for the structure functions, one finds

$$A \simeq 0.69, \qquad C \simeq 0.06, \qquad D \simeq 2.4 \quad .$$

This is all that is needed for the description of the deep inelastic scattering.

One obtains the results summarized in the Figures 1, 2, 3, 4, 5, concerning the unpolarized e-N scattering structure functions, the asymmetry in the same process, and the two independent unpolarized structure functions of $\nu$-nucleon scattering. They are compared, whenever possible, with the experimental data, and we have an overall satisfying agreement. The various sum rules which combine electron, neutrino and antineutrino data,are also well satisfied (for details, see (15)).

It appears that the behavior of $R^{np}(x)$ and the asymmetries $A^{ep}(x)$ and $A^{en}(x)$, is mainly controlled by the mixing, except at low $x$. The gluons control the height and position of the maximum of $F_2^{ep}(x)$, together with the radius of the wave function $R^2$. The pairs control the low $x$ behavior.

In the static and low $t$ region, $G_A/G_V$ is controlled by the value of $\mu_q^2/R^2$ (the small components effect (i)), while the neutron charge radius is essentially fixed by the mixing (ii), it does not depend much on the small components effect.

For any further detail, we refer to the paper[15].

## VI. How far is the distinction between constituent and current quarks relevant ?

For at least three years, one has been constantly speaking of a distinction between two kinds of quarks, current and constituent, which should be essential in clarifying the dynamics of quarks and hadrons. We wonder whether, within the conceptual framework of a realistic quark model, this distinction is so clarifying as has been claimed, or whether it should not be replaced by much more simple-minded notions of (relati-

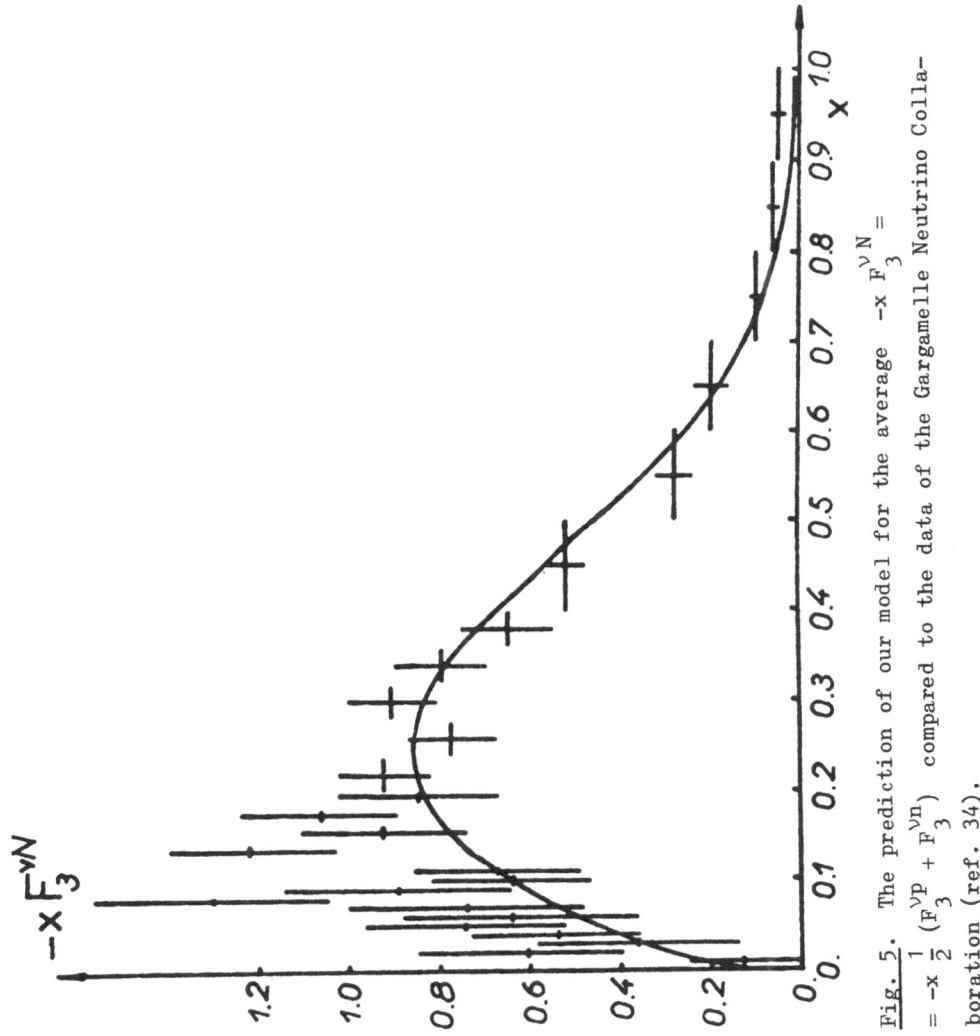

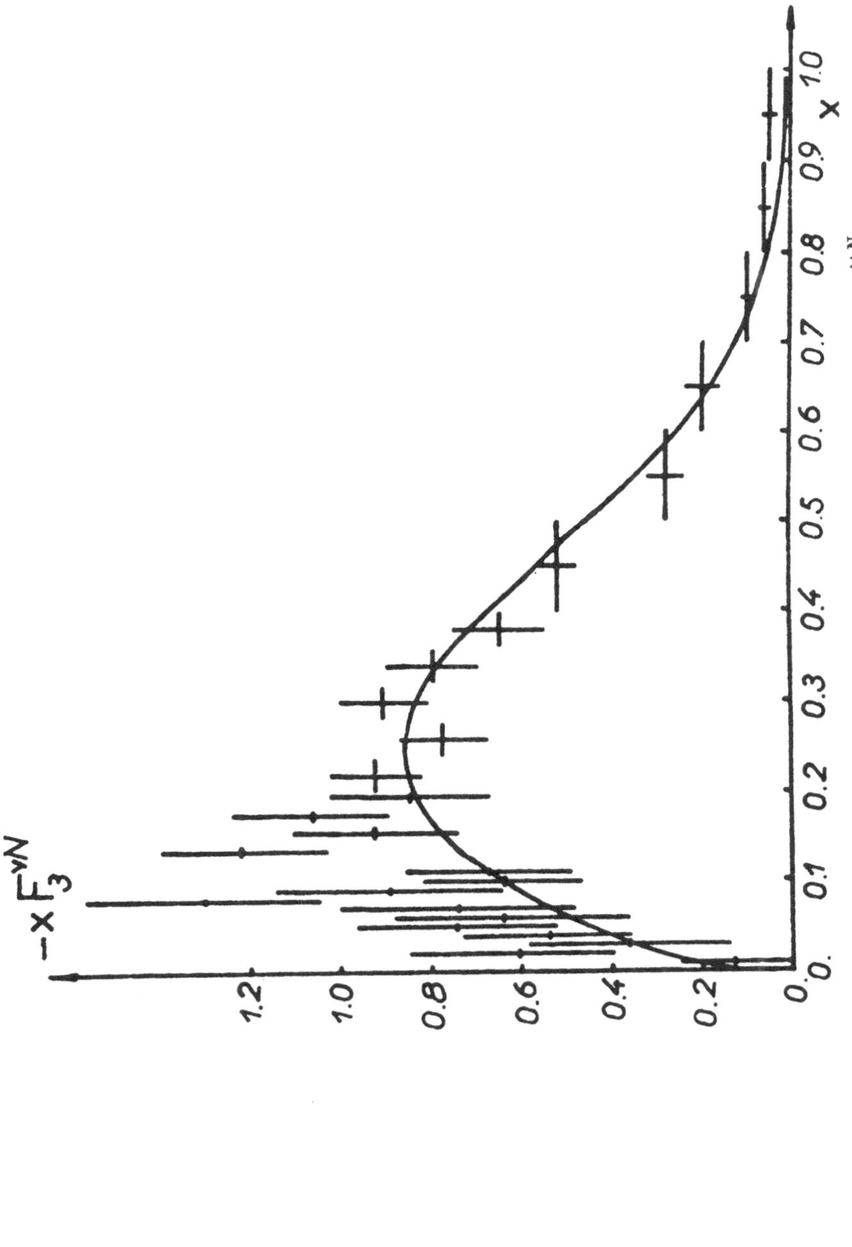


Fig. 5. The prediction of our model for the average $-x\, F_3^{\nu N} = -x\, \frac{1}{2}\, (F_3^{\nu p} + F_3^{\nu n})$ compared to the data of the Gargamelle Neutrino Collaboration (ref. 34).

vistic) quantum mechanics. In fact, a critical discussion of the current and constituent quarks philosophy is useful, at the end of our analysis, to emphasize our particular approach of the deep inelastic scattering.

A. First, one should notice that, as far as we understand the work of Melosh, the real distinction is between two Hilbert bases, where respectively the current operator or the wave function of the hadronic state are known and simple. Why should the existence of these two Hilbert bases be assigned to the existence of two kinds of particles ? As it has nothing to do with what is commonly meant by two kinds of particles (e.g. leptons and baryons) we do not find the expression much clarifying. And it appears still less clarifying when one realizes that, in a subsequent work of Melosh[8], the phenomenon reduces to a Wigner rotation, i.e. to a simple kinematical transformation (as we have independently suggested[6]).

More generally, when one speaks of "current" quarks, "constituent" quarks, "dual" quarks, one really means only a <u>subjective</u> distinction between various theories or approaches[*] which all use some concept of quark. But the terminology is misleading, because it suggests an objective distinction between various kinds of quarks, which is quite another story. The suggestion could even be wrong in some cases.

For our part, we insist that there should be fundamentally <u>one kind</u> of quarks, and that one should try to discover the link between the various approaches by insisting on this fundamental identity (but also by recognizing the most various aspects that quarks may present in various processes).

B. A more specific criticism comes from the present analysis of the deep inelastic scattering.

What is really a problem is not that we observed several kinds of quarks, but that we observed a departure from the naive SU(6) results

---

[*] i.e. : The current algebra, the naive quark model, the duality diagrams.

based on the naive quark model. Now, why is the naive quark model only an approximation ? There are at least three distinct reasons :

(i) The internal quark motion inside the hadron is relativistic and one should take it into account with Dirac small components.

(ii) The nucleon is not a pure 56, but a mixing with a $(\underline{70},0^+)$ is required.

(iii) The quarks are dressed by virtual particles, such as gluons and $q\bar{q}$ pairs.

The only effect in which one could recognize something of a distinction between two kinds of quarks (current and constituent) is the third one. The bare quarks, which have a point-like current structure, would be the current quarks, while the dressed quarks, in terms of which the structure of the hadronic states is simple : $q\bar{q}$ or 3q, would be the constituent quarks. This is the interpretation of Altarelli et al.(14). But precisely it is not the effect considered by Melosh, who considered rather the effect (i). And precisely, on the basis of our analysis, we suggest that the departure from SU(6) is not due to (iii) (quark dressing) but to (i) and (ii).

Now what is the point in speaking of a distinction between two kinds of quarks for the first two effects (i) and (ii) ? They do not concern the quarks by themselves, they are complications of the hadron state due to the relative motion or interaction of the quarks with respect to each other. Last but not least, while one invokes Melosh who is concerned by (i), the main effect at large $x$ is in our mind (ii).

Concerning the axial couplings, we have shown that (i) is the main effect since it yields the dominant SU(3) ⊗ SU(3) representations of the nucleon wave function(6), and even the mixing parameters determined phenomenologically by Buccella et al.(35)

We do not mean however that the dressing effect (iii) could not play an important role. If there seems to be a quark anomalous magnetic moment, why should $g_A$ be equal to 1 ?

We have assumed $g_A = 1$ only for the sake of simplicity and because we had no means to estimate the "radiative" corrections to the quark axial coupling.

Finally, the three effects could be included in a generalized transformation between current and constituent quarks. But at this point, it would be simply a definition. Perhaps, later, one will arrive at a comprehensive understanding of the various effects involved on the basis of a single law of interaction between the quarks, but it still requires much effort ! Meanwhile, we prefer to pursue the analysis with the help of as many as possible simple-minded concepts taken from the relativistic quantum mechanics, such as the effects (i) to (iii).

## References


(1) F. Gürsey and L.A. Radicati, Phys. Rev. Letters 13, 173 (1964); A. Pais, Phys. Rev. Letters 13, 175 (1964).

(2) R. Dashen and M. Gell-Mann, Third Coral Gables Conference on Symmetry Principles at High Energies, 1966 (Freeman, 1966), p. 168;

(3) For a review, see for instance J. Rosner, Physics Reports C 11, 189 (1974).

(4) E. Hirsch, V. Karshon, H.J. Lipkin, Phys. Letters 36B, 385 (1971).

(5) J.I. Friedman and H.W. Kendall, Ann. Review of Nucl. Science 22, 203 (1972).

(6) A. Le Yaouanc, L. Oliver, O. Pène and J.C. Raynal, Phys. Rev. D9, 2636 (1974).

(7) L. Micu, Nucl. Phys. B10, 521 (1969) ; A. Le Yaouanc, L. Oliver, O. Pène and J.C. Raynal, Phys. Rev. D8, 2223 (1974); D9, 1415 (1974); D 11, 1272 (1975).

(8) H.J. Melosh, Phys. Rev. D9, 1095 (1974) ; A.J.G. Hey, P.J. Lichtfield and R.J. Cashmore, Nucl. Phys. B 95, 516 (1975). For a review, see ref. 3.

(9) E. Colglazier and J. Rosner, Nucl. Phys. B27, 349 (1971); W. Petersen and J. Rosner, Phys. Rev. D 6, 820 (1972) ; D. Faiman and J. Rosner, Phys. Letters 45B , 357 (1973) ; D. Faiman and D. Plane, Phys. Letters 39B, 358 (1972), Nucl. Phys. B50, 379 (1972).

(10) A.N. Mitra and M. Ross, Phys. Rev. 158, 1630 (1967).

(11) F.E. Close, Phys. Letters 43B, 422 (1973).

(12) R.P. Feynman, High Energy Collisions (ed. by C.N. Yang _et al._, Gordon and Breach, New York, 1969), p. 237.

(13) F.E. Close, H. Osborn and A.M. Thomson, Nucl. Phys. B77, 281 (1974). F.E. Close, Nucl. Phys. B80, 269 (1974).

(14) G. Altarelli, N. Cabibbo, L. Maiani and R. Petronzio, Nucl. Phys. B 69, 531 (1974).

(15) A. Le Yaouanc, L. Oliver, O. Pène and J.C. Raynal, preprint LPTHE 75/11, to appear in Phys. Rev. D (1975).

(16) R.H. Dalitz, in High Energy Physics, Ecole d'Eté de Physique Théorique, C. DeWitt and M. Jacob, Eds., Les Houches, 1965 (Gordon and Breach, New York, 1966) ; XIIIth International Conference on High-Energy Physics, Berkeley, 1966.

(17) G. Morpurgo, Physics 2, 95 (1965).

(18) D. Faiman and A.W. Hendry, Phys. Rev. 173, 1720 (1963) ; 180, 1572 (1968) ; 180, 1609 (1969).

(19) J.D. Bjorken and E.A. Paschos, Phys. Rev. 185, 1975 (1969).

(20) J. Kuti and V.F. Weisskopf, Phys. Rev. D4, 3418 (1971).

(21) A.L. Licht and A. Pagnamenta, Phys. Rev. D2, 1150 (1970).

(22) A. Le Yaouanc, L. Oliver, O. Pène, and J.C. Raynal, Nucl. Phys. B37, 552 (1972) ; with P. Andreadis and A. Baltas : Annals of Physics 88, 242 (1974).

(23) A. Le Yaouanc, L. Oliver, O. Pène and J.C. Raynal, Phys. Rev. D 11, 680 (1975).

(24) P.N. Bogolioubov, Ann. Inst. Henri Poincaré 8, 163 (1968).

(25) B.H. Kellett, Phys. Rev. D10, 2269 (1974).

(26) S. Kitakado, IXth Rencontre de Moriond (1974), ed. J. Tran Thanh Van.

(27) J.D. Bjorken, Phys. Rev. D1, 1376 (1970).

(28) J. Ellis and R. Jaffe, Phys. Rev. D9, 1444 (1974).

(29) L.M. Sehgal, Phys. Rev. D10, 1663 (1974).

(30) This point has been emphasized by F.E. Close (ref. 11) and independently by M. Chaichian and S. Kitakado, Nucl. Phys. B59, 285 (1973).

(31) V.E. Krohn and G.R. Ringo, Phys. Rev. D8, 1305 (1973).

(32) A. Le Yaouanc, L. Oliver, O. Pène and J.C. Raynal, preprint LPTHE 72/6 (unpublished).

(33) G. Miller et al., Phys. Rev. D5, 528 (1972).

(34) H. Deden et al., Gargamelle Neutrino Collaboration, Nucl. Phys. B85, 269 (1975).

(35) F. Buccella, M. De Maria and L. Lusignoli, Nucl. Phys. B6, 430 (1968).

## Acknowledgements

The authors are indebted to the organizing committee, in particular to Professor G. Morpurgo, for their kind invitation to participate in the Workshop.

# RELATIVISTIC MOTION OF COMPOSITE SYSTEMS WITH NONRELATIVISTIC INTERNAL DYNAMICS

G. Morpurgo

Istituto di Fisica dell' Università-Genova
Istituto Nazionale di Fisica Nucleare
Sezione di Genova

1. This report is essentially a summary of two lectures given here at last year's School of Subnuclear Physics.[1] For details I refer the reader to these lectures and confine myself here only to a general idea of what I tried to do. It should be stressed that this attempt is, at best, only a preliminary step, interesting, if at all, only to show clearly where the difficulties come from when, in a relativistic problem, one tries to separate the external kinematics from the internal dynamics.

Clearly, at the present stage, field theory is the only instrument we have to guarantee the relativistic invariance of a calculation. However, field theory is afflicted by well-known difficulties, especially in the treatment of bound states. This is the reason why attempts have been made for many years to construct a relativistic dynamics of composite systems in which only the degrees of freedom of the bound particles, and not those of the field, intervene.[2]

What I will summarize here belongs to this line of investigation, but is very different from the work done so far[2] both in its aim and in procedure.

To start as simply as possible, consider an example: A harmonic oscillator consists of two harmonically bound particles of the same mass. One of the particles is charged. We assume that the relative motion of the two particles is nonrelativistic; in other words, in the rest frame - in which the total momentum of our two-body system is zero - we have to deal with a standard nonrelativistic system of quantum mechanics. There is no problem in calculating, for instance, the lifetime of the radiative decay of the oscillator in a transition between two states: $a \to b + \gamma$; call the result $\tau_0$. If the oscillator is now moving with velocity v relative to us, we know that the lifetime becomes ($\beta = v/c$):

$$\tau = \tau_0/\sqrt{1 - \beta^2} \tag{1}$$

The exercise which we propose is as follows: Try to calculate directly, that is, without making use of equation (1), the lifetime in the reference frame in which the oscillator is moving. As we shall see, this is not so easy. Indeed, to perform this exercise, we must be able to construct the Hamiltonian of our harmonic oscillator in fast motion and in interaction with the electromagnetic field.

2. Before discussing how, and under what limitations, the construction of the above Hamiltonian can be carried out, we mention a few physical problems which stimulated this study.

A. Consider the production of "positronium" by a high-energy $\gamma$ ray in the Coulomb field of a heavy nucleus. If $E_\gamma$ is sufficiently high, the positronium which is produced moves relativistically. To calculate the rate of this process - in the impulse approximation - we must know the wave function of such a moving positronium. This problem is similar to that of the harmonic oscillator stated previously in the sense that there also to calculate the lifetime directly one has to know the wave functions of the oscillator in motion. One could object that in both cases there is no problem, that in fact one should simply use the Bethe-Salpeter equation. This is (at least for the case of positronium) certainly true. Our point, however, is that, in the rest frame, the two problems described are ordinary problems of nonrelativistic quantum mechanics, and one should be able to solve them without invoking the apparatus of field theory which underlies the BS equation. Stated differently, we are confronted with two problems which one would like to call problems of pure kinematics; the difficulty is that, in relativistic mechanics, kinematics and internal dynamics are not easily separable.

B. Another similar class of processes is that of the reactions between nuclei in relativistic motion. Nuclei have essentially nonrelativistic internal dynamics, and, again, one would like to have an instrument to describe at least those relativistic processes involving nuclei in which production of pions or antinucleons does not take place.

C. A class of "nuclear" reactions of particular interest is represented by some processes in the nonrelativistic quark model. Indeed, my interest in this kind of problems was enhanced precisely by some problems in this field.[3] Consider, for example, the

$\omega \to \pi\gamma$ decay: a $q\bar{q}$ state (the $\omega$) decays into another $q\bar{q}$ state (the $\pi$) plus a photon. Because the pion is much lighter than the $\omega$, it moves relativistically after the decay. To compute the rate of the process we should know its wave function. The fact that in the particular case of the $\omega \to \pi\gamma$ decay (and, more generally, of $V \to P\gamma$ decays) the difficulty stated above can in fact be eliminated - as done in Ref. 4 and explained in detail in Ref. 1 - is in a sense irrelevant, because in general the difficulty remains.

3. Note that the $V \to P\gamma$ decays just mentioned are similar to the $a \to b\gamma$ decay of the harmonic oscillator except for an important point: in the case of the harmonic oscillator, the condition

$$(m_a - m_b)/m_a \ll 1 \tag{2}$$

is satisfied whereas in the actual $V \to P\gamma$ decays it is not. The treatment which we are going to give will apply to the problem of the harmonic oscillator only under the condition (2). It must be stressed that it will therefore not apply, for the moment, to problems like that of the $V \to P\gamma$ transitions.

4. We refer from now on to the harmonic oscillator, and proceed to write the Hamiltonian of a freely moving harmonic oscillator composed of two particles, 1 and 2, harmonically bound. We assume, for simplicity, that the two particles are spinless. Call $\underset{\sim}{x}_1$ and $\underset{\sim}{x}_2$ the position vectors of the two particles, $\underset{\sim}{X} = (\underset{\sim}{x}_1 + \underset{\sim}{x}_2)/2$ and $\underset{\sim}{x} = \underset{\sim}{x}_1 - \underset{\sim}{x}_2$, and introduce the variable $\underset{\sim}{\xi}$ defined by

$$\underset{\sim}{\xi} = \underset{\sim}{x} + (\underset{\sim}{x} \cdot \dot{\underset{\sim}{X}})\dot{\underset{\sim}{X}}\tilde{f}(\dot{\underset{\sim}{X}}^2) \tag{3}$$

where we have set $c = 1$ and $\dot{\underset{\sim}{X}}$, the overall velocity of the oscillator, is a time-independent vector because the oscillator is free.

In (3) we have put

$$\tilde{f}(\dot{\underset{\sim}{X}}^2) = \frac{1}{\dot{\underset{\sim}{X}}^2}\left(\frac{1}{\sqrt{1 - \dot{\underset{\sim}{X}}^2}} - 1\right) \tag{4}$$

Note that

$$\xi_{\parallel} = x_{\parallel} / \sqrt{1 - \dot{\underset{\sim}{X}}^2}, \quad \xi_{\perp} = x_{\perp}$$

where longitudinal and transverse refer to the direction of $\dot{\underset{\sim}{X}}$.

The assumption that the internal motion of the oscillator is nonrelativistic can be written

$$A\omega/c \ll 1 \tag{5}$$

where A is the amplitude of oscillation in the rest system. Equation (5) states that the time A/c spent in the rest system by a light signal in going from particle 1 to particle 2 is much smaller than the period $1/\omega$ of the oscillator. Under the assumption (5) a particularly simple (although approximate) description of the motion of the oscillator is obtained - for any value $\dot{\underset{\sim}{X}}$ of its overall speed - by considering the positions of the two particles at the same time t. This is the time which will appear in the following equations; it should be stressed that the description which we are going to give is approximate in the sense that terms of higher order in the quantity (5) are neglected.

In this situation, explained in great detail in Ref. 1, the Hamiltonian of the freely moving oscillator is shown to be

$$H^{(o)} = \sqrt{\underset{\sim}{P}^2 + (M + \underset{\sim}{p}^2/2\mu + k\underset{\sim}{\xi}^2/2)^2} = \sqrt{\underset{\sim}{P}^2 + h^{(o)2}} \tag{6}$$

where we have put

$$h^{(o)} = M + \underset{\sim}{p}^2/2\mu + k\xi^2/2 \tag{7}$$

It is possible to check, by writing the equations of motion corresponding to (6), that we get the correct approximate equations. Note in expression (6) the presence of the variable $\underset{\sim}{\xi}$ defined in (3); it is only because we have introduced this variable, instead of $\underset{\sim}{x}$, that we could obtain a separable Hamiltonian, that is, a Hamiltonian of the form $\sqrt{\underset{\sim}{P}^2 + h^{(o)2}}$, where $\underset{\sim}{P}$ is the total momentum and $h^{(o)}$ depends on the internal variables only. This is satisfactory, in the sense that we have achieved our first aim of describing the motion of the free harmonic oscillator by means of a Hamiltonian, independently of how high its translational speed $\dot{\underset{\sim}{X}} = \underset{\sim}{P}/H^{(o)}$ is.

When the above Hamiltonian is translated into a quantum-mechanical operator, the wave functions of our system can be written as

$$\phi^{(o)}_{\underset{\sim}{P},\alpha}(\underset{\sim}{X},\underset{\sim}{\xi},t) = N_{\underset{\sim}{P},\alpha} \exp\{i\underset{\sim}{P} \cdot \underset{\sim}{X}\}\phi_\alpha(\underset{\sim}{\xi})\exp\{-iE_{\underset{\sim}{P},\alpha}t\} \tag{8}$$

with

$$E_{\underset{\sim}{P},\alpha} = \sqrt{\underset{\sim}{P}^2 + m_\alpha^2}$$

$m_\alpha$ being the rest mass of the oscillator in the state $\alpha$; of course, $h^{(o)}\phi_\alpha(\underset{\sim}{\xi}) = m_\alpha\phi_\alpha(\underset{\sim}{\xi})$.

5. So far so good. But how do we write the Hamiltonian in the more interesting case in which the harmonic oscillator is coupled to external forces? A case of some interest is that of an electromagnetic field interacting with the charged particle (say particle 1) of the harmonic oscillator. One might be tempted in this case to proceed as follows. (a) Write the unperturbed Lagrangian corresponding to (6). As is easily seen, this is

$$L^{(o)} = -\sqrt{1 - \dot{\underset{\sim}{X}}^2}\left(M + k\underset{\sim}{\xi}^2/2 - \frac{\mu\, \dot{\underset{\sim}{\xi}}^2}{2(1 - \dot{\underset{\sim}{X}}^2)}\right) \tag{9}$$

(b) Write the usual electromagnetic interaction, adding to $L^{(o)}$ the expression $L_{e.m.} = e\dot{\underset{\sim}{x}}_1 \cdot \underset{\sim}{A}(\underset{\sim}{x}_1) - e\phi(\underset{\sim}{x}_1)$ and expressing $\underset{\sim}{x}_1$ in terms of $\underset{\sim}{\xi}$.

This procedure is however not correct, for the following reason: The relationship between $\underset{\sim}{x}_1$ and $\underset{\sim}{\xi}$ contains the derivative $\dot{\underset{\sim}{X}}$; it is in fact

$$\underset{\sim}{x}_1 = \underset{\sim}{X} + \frac{1}{2}[\underset{\sim}{\xi} + (\underset{\sim}{\xi} \cdot \dot{\underset{\sim}{X}})\, \dot{\underset{\sim}{X}}\, f(\dot{\underset{\sim}{X}}^2)] \tag{10}$$

where

$$f(\dot{\underset{\sim}{X}}^2) = \frac{1}{\dot{\underset{\sim}{X}}^2}\left(\sqrt{1 - \dot{\underset{\sim}{X}}^2} - 1\right)$$

Therefore $L_{e.m.}$, when expressed in terms of $\underset{\sim}{\xi}$ and $\underset{\sim}{X}$, contains $\ddot{\underset{\sim}{X}}$, and the total Lagrangian contains derivatives of higher than first order, which is not acceptable for a Lagrangian. Stated differently, the transformation (10) is not a transformation between Lagrangian coordinates.

Although we have not yet studied what to do in general (we expect that one should use new, interaction-dependent variables replacing $\underset{\sim}{\xi}$ and $\underset{\sim}{X}$ from the start), there is however one particular case in which we can proceed. This is the case where we confine ourselves to consideration of all the effects of the electromagnetic interaction only to first order in e. In that case the addition of Le.m. to $L^{(o)}$ does not introduce terms in $\ddot{\underset{\sim}{X}}$ because such terms are necessarily of second order in e (since $\dot{\underset{\sim}{X}}$ changes only due to coupling with the electromagnetic field), and we can write, up to the first order in e, $L = L^{(o)} + L^{(1)}$, with

$$L^{(1)} = e\underset{\sim}{A}(\underset{\sim}{X} + \frac{1}{2}[\underset{\sim}{\xi} + (\underset{\sim}{\xi} \cdot \dot{\underset{\sim}{X}})\dot{\underset{\sim}{X}}f(\dot{\underset{\sim}{X}}^2)]) \cdot \{\dot{\underset{\sim}{X}} + \frac{1}{2}[\dot{\underset{\sim}{\xi}} + (\dot{\underset{\sim}{\xi}} \cdot \dot{\underset{\sim}{X}})\dot{\underset{\sim}{X}}f(\dot{\underset{\sim}{X}}^2)]\} \tag{11}$$

where we have omitted the term in the scalar e.m. potential $\phi$, because we shall not need it in what follows.

From the above expression of L the Hamiltonian can be written up to first order in e: $H = H^{(o)} + V$, where $H^{(o)}$ is given by (6) and V is defined by

$$VH^{(o)} + H^{(o)}V = -e[2\underset{\sim}{P} + \{\frac{h^{(o)}}{\mu}\underset{\sim}{p} + \frac{h^{(o)}}{\mu}(\underset{\sim}{p}\cdot\underset{\sim}{P})\underset{\sim}{P}\overline{f}(h^{(o)},\underset{\sim}{P}^2)\}]\cdot A(\ldots) \quad (12)$$

where

$$\underset{\sim}{A}(\ldots) = \underset{\sim}{A}(\underset{\sim}{X} + \tfrac{1}{2}\{(\underset{\sim}{\xi}\cdot\underset{\sim}{P})\underset{\sim}{P}\overline{f}(h^{(o)},\underset{\sim}{P}^2) + \underset{\sim}{\xi}\})$$

and

$$\overline{f} = H^{(o)2}/P^2(h^{(o)}/H^{(o)} - 1)$$

It is obvious that whereas the structure of $H^{(o)}$ is simple in terms of $\underset{\sim}{\xi}$ and its conjugate momentum $\underset{\sim}{p}$, the structure of the electromagnetic interaction, expressed in the same variables, is complicated. One can thus begin to appreciate, at least in a particular case, why the separation of internal and external variables is so complicated in the relativistic case.

6. It is possible, in a particularly simple situation, to check the above treatment by calculating - using the formulas just given - the electromagnetic decay $a \to b + \gamma$ for an oscillator in motion and comparing the result with the exact covariant vertex; if the two agree we may be sure that equation (1), in particular, is satisfied. To be more definite, consider a decay of the type $V \to S + \gamma$, where V is a vector meson and S a scalar meson. We compare the fully covariant matrix element of this decay (which is unique, except for a form factor, because only one possible covariant vertex exists) with the matrix element of the electromagnetic interaction (12) between a state of our harmonic oscillator with L = 1 (vector meson) and a state with L = 0 (scalar meson).

The covariant vertex for the decay $V \to S + \gamma$ is

$$\mathcal{F}_{\alpha\beta}\partial_\beta A_\alpha S \quad (13)$$

where $\mathcal{F}_{\alpha\beta} = \partial_\alpha V_\beta - \partial_\beta V_\beta$ and S and $A_\alpha$ are, respectively, the scalar and electromagnetic fields. The matrix element G arising from the above vertex (13) is, omitting a factor $C(E_V E_S k)^{-1/2}$ where C is a numerical constant

$$G \propto (\varepsilon P^a)(\eta k) - (kP^a)(\varepsilon\eta) \quad (14)$$

Here $P^a$, $\eta$ are the four momentum and four polarization of the vector meson and k, $\varepsilon$ are those of the photon. Expression (14) can be readily transformed, in three-dimensional notation, into

$$G \propto (\underset{\sim}{\varepsilon} \cdot \underset{\sim}{P}^a)\left(\underset{\sim}{\eta}^{(P)} \cdot \underset{\sim}{k} - \frac{\underset{\sim}{\eta}^{(P)} \cdot \underset{\sim}{P}^a}{E^{(a)}} k\right) - (\underset{\sim}{k} \cdot \underset{\sim}{P}^a - kE^a)(\underset{\sim}{\varepsilon} \cdot \underset{\sim}{\eta}^{(P)}) \quad (15)$$

We come now to the calculation of the matrix element $a \to b + \gamma$ starting with the unperturbed Hamiltonian (6) and using the perturbation (12). This is rather complicated. It is performed in detail in Ref. 1, and here we can only give the results, stating the conditions under which they were deduced.

The essential condition is expressed by the inequality (2). Indeed, only if this condition is imposed will some problems of quantum-mechanical ordering of operators which arise when the classical Hamiltonian (6) + (12) is transformed into an operator disappear. As already stated, the necessity of imposing this condition precludes, for the moment, the possibility of applying these results to cases such as that of the $\omega \to \pi\gamma$ decay.

If the limitation (2) is satisfied we can, as shown in Ref. 1, reproduce with our technique the result (15) [as well as the multiplying factor $(E_V E_S k)^{-1/2}$ omitted in writing (15)]. The correspondence between the matrix elements of the oscillator and the polarization $\eta^{(P)}$ turns out to be as follows. Define

$$\eta^{(o)} = \langle b|\underset{\sim}{\xi}|a\rangle/\alpha^{(o)}$$

where $\langle b|$ and $|a\rangle$ are the internal wave functions of the states of the oscillator with L = 1 and L = 0 and $\alpha^{(o)}$ is a scalar quantity (a number) constructed in terms of these wave functions. Define also $\underset{\sim}{\eta}^{(P)}$ by

$$\underset{\sim}{\eta}^{(o)} = \underset{\sim}{\eta}^{(P)} + \underset{\sim}{P}\,(\underset{\sim}{\eta}^{(P)} \cdot \underset{\sim}{P})\overline{f}$$

With this definition of $\underset{\sim}{\eta}^{(P)}$ the result (15) is reproduced. As already stated, the calculation is rather lenghty, and we cannot give the details here.

7. For a complete presentation of the conclusions we must refer again to Ref. 1 (final section). Only one point will be mentioned here. It appears clearly from the above treatment, in spite of all its limitations, that once it is assumed that the internal motion is nonrelativistic, the velocity of the composite system with respect to us can be as large as we like. This remark shows that the approach presented here is fundamentally different from those listed in Ref. 2, where everything is expanded in series

of v/c, without distinction in this respect between internal and translational velocity. But in fact, the distinction between the present approach and that of Ref. 2 is much greater than this. The point is that, contrary to the approach of Ref. 2, we accept passively the fact that there exist motions of composite systems which, to a very good approximation, can be parametrized or described, in their rest system, in terms of nonrelativistic equations of motion. This implies that we are not, so to speak, interested at this stage in the fact that in a pendulum or positronium atom one should, to be exact, take into account, the effects due to the finite velocity c at which the actions between the constituents propagate. We are only interested in the problem of being able to treat by ordinary quantum mechanics a moving composite system the internal dynamics of which is specified in its rest system. This is what we have exemplified, under many restrictions, in the problem just considered.

References

1. G. Morpurgo, "Lepton and Hadron Structure," A. Zichichi, ed., Academic Press, 1975 (Erice lectures 1974).

2. L. L. Foldy, Phys. Rev. 122, 289 (1961); R. A. Krajcik and L. L. Foldy, Phys. Rev. Letters 24, 545 (1970); M. K. Liou, Phys. Rev. D9, 1091 (1974); F. Close and H. Osborn, Phys. Rev. D2, 2127 (1970). These last two papers contain many other references.

3. G. Morpurgo, Physics 2, 95 (1965).

4. C. Becchi and G. Morpurgo, Phys. Rev. 140B, 687 (1965).

# DYNAMICS OF THE MESON SPECTRUM

Hans Joos
Deutsches Elektronen-Synchrotron DESY, Hamburg

## 1. INTRODUCTION

In a series of papers [1], M. Böhm, M. Krammer and myself have made an attempt of a dynamical interpretation of the mesons and their properties in the framework of a relativistic quark model. Being aware of the scarcety [2] of established principles of quark dynamics, we pursued the following line:

In order to incorporate the general principles of physics, we use the framework of general field theory. Hence it seemed natural to base our model on a general Bethe-Salpeter equation.

We wanted to take into account the negative result of the search for quarks. For this we assumed a large quark mass as a mean of practical quark confinement. Our approach should be able to absorb phenomenological experience. Therefore, besides being simple at the beginning, our ansatz was planned to be general and flexible.

It was our hope that from a model developed along these lines and applied to many different processes, we could learn what are important dynamical parameters and what are their qualitative relations. Such an insight should help to bridge over the gap between a fundamental theory and purely phenomenological explorations.

The result of our work was described in a concise form at the Erice work shop on Quarks 1974 [3]. There we presented a coherent dynamical interpretation of the meson spectrum, the strong and leptonic decays of mesons, and of meson form factors, which lead to a satisfactory numerical agreement with the experimental values.

Since then the discovery of the new particles [4] exercised its strong influence on elementary particle physics. This and some new theoretical developments changed also my attitude towards my lectures at this workshop. So I shall not elaborate further on the applications of our model to "old" mesons; there is nothing about mesons to add to the published results [5]. Instead of that I would like first to discuss how the model looks in view of the new particles.

There are also very interesting new theoretical approaches to the problem of quark confinement and strongly bound systems, like the field theoretical description of bags [6]. How do these new ideas compare with our field theoretical, phenomenological approach? My interest in this question is twofold. First we might be able to derive a justification of some of our phenomenological assumptions from more fundamental equations. On the other hand I am interested to see, how these new models cope with the many solved and unsolved problems which we met in our various applications. With this in mind, I shall give a critical review of the assumption underlying the relativistic heavy quark model, and I shall compare these with some formal aspects of the SLAC-bag model. Further, I want to rise the question of chiral symmetry and PCAC.

For the most part these comments do not contain solid new results, but are meant mainly to stimulate the discussion at this work shop.

## 2. DYNAMICAL RELATIONS

Let me first explain what I consider as "the dynamics of the meson-spectrum".

The mesons and their properties are described with help of certain dynamical constants, which have dimensions - in contrast to symmetry coefficients like Clebsch-Gordan coefficients etc. Let me mention some of the most important ones:

(a) The MESON SPECTRUM is determined by the Regge slope $\alpha'$ and the Regge-intercept $\alpha_0$ :

$$M^2 = \alpha'^{-1} j + \alpha_0 \tag{1}$$

j spin, M - mass of the resonances.

(b) The relation between STRONG DECAYS of different particles on a Regge-trajectory is mainly governed by a RANGE-PARAMETER a, which appears in the threshold factor $B_j$ in the well-known phenomenological decay formula

$$\Gamma = g^2 \left[CG(SU(3))\right]^2 B_j(P_{cm}) \left(\frac{M_0}{M_R}\right)^2 P_{cm} \tag{2}$$

with

$$B_j = \frac{2^j}{(2j+1)!!} (a P_{cm})^{2j} F(a P_m) \tag{2'}$$

(c) The LEPTONIC DECAYS, like $\rho^0, \omega, \phi \to e^+e^-$, $\pi \to \mu\nu$ are calculated from matrix elements

$$\langle 0 | j_\mu(0) | V \rangle = g_V \, \varepsilon_\mu^{s_3}(P) \tag{3}$$

where $\varepsilon_\mu$ denotes the kinematical polarization vector and the "density value" $g_V$ is the dynamical parameter.

It is the main problem of the dynamics of the meson spectrum to find relations between such dynamical parameters, which by their own nature are not of symmetry type. Of course, it is also required that the dynamics are compatible with the phenomenological picture on which the introduction of these parameters is based.

In our model the dynamics are based on the calculation of B.S.-amplitudes

$$\chi(q,P) = (2\pi)^{3/2} \int dx \, e^{iqx} \langle 0 | T \psi(\tfrac{x}{2}) \bar{\psi}(-\tfrac{x}{2}) | Mes \rangle$$

$= $ 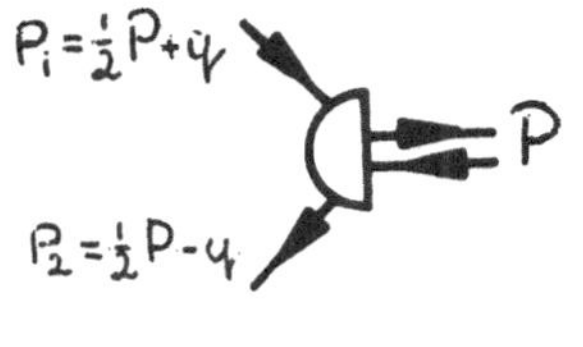


(4)

from a phenomenological B.S. equation.

There are solutions for states lying on linear Regge-trajectories: $M^2 = \alpha'^{-1}(n+2r) + \alpha_0$, $j = n, n-2, \cdots$, $r = 0, 1, 2 \ldots$ which have the spin-singlet-triplet structure of the non-relativistic quark model. From the B.S.-amplitudes we can calculate directly the leptonic decay matrix elements with help of the current expression $j_\mu(x) = Z\bar{\Psi}(x) Q \gamma_\mu \Psi(x)$.

$$\langle 0|j_\mu(0)|v\rangle = (2\pi)^{-3/2} Z \,\mathrm{Trace}\left(Q\gamma_\mu \int dq\, X(q,P)\right) \tag{5}$$

The hadronic coupling constants are calculated approximately by the evaluation of the triangle graph

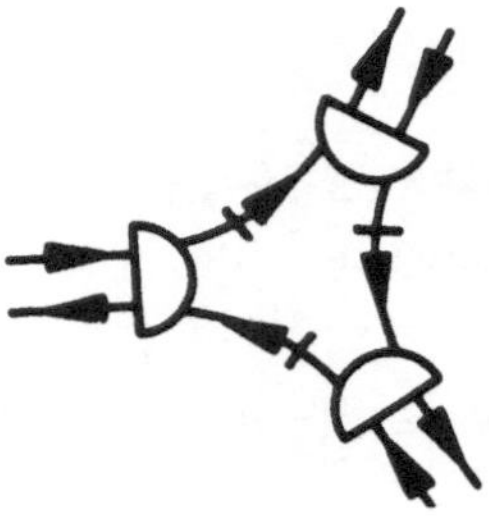

Fig. 1.

By this procedure, which is described in detail in Ref. 3, we get the dynamical relations

$$a \approx 1.2\sqrt{\alpha'} \tag{6}$$

expressing the range parameter of hadronic decays by the Regge slope, and

$$g_V \approx 0.18 (-1)^r Z \frac{\sqrt{r+1}}{\alpha'} \langle Q_V \rangle \tag{7}$$

for the leptonic decay constants of the radially excited vector mesons; $g_V = 0$ for the other excited vector mesons. $\langle Q_V \rangle$ denotes the quark charge factor $1/\sqrt{2}$, $1/3\sqrt{2}$, $-1/3, \ldots$ for $\rho^0, \omega, \phi, \ldots$ , Z the current renormalization constant.

In order to illustrate the significance of these relations, I mention only two applications. It follows from Eq. (2), that a determines mainly the relation between partial widths of decays, which differ only in the spin of the decaying resonance. Hence the following example from our calculations with Eq. (6):

| A | $\rho(770)$ | f(1270) | g(1680) |
|---|---|---|---|
| $j^{\pi g}$ | $1^{-+}$ | $2^{+-}$ | $3^{-+}$ |
| $\Gamma(A \to 2\pi)$ exp. [7] | 146 | 130 | 64 MeV |
| theor. | input | 140 | 76 MeV |

From Eq. (7) we get $g(\rho(770))/g(\rho'(1600)) = 1/\sqrt{2}$ to be compared with the latest experimental value [8] $g(\rho)/g(\rho') = 0.73 \pm 0.15$.

This type of results let us conclude that our model reflects correctly some of the dynamical features of the old mesons. Now let us consider the new ones.

## 3. SOME REMARKS ON THE DYNAMICS OF THE NEW PARTICLES

Because of the unknown "flavour" of the new particles, comments on their dynamics are pure speculations. But we follow the general trend, assume the charm model, and discuss how the new particles show the characteristic features of the dynamics of the meson spectrum discussed above.

Let us first have a look at the singlet-triplet structure and the mass distribution of the spectrum. In Table I we compare the series of mesons with isospin I = 1, i.e. quark configuration $\bar{p}p - \bar{n}n$, with the observed new particles [9] considered as hidden charm states with quark configuration $\bar{c}c$. Similarities in the mass distributions give some appeal to this association of dynamical quark quantum numbers to the new particles. But if an oscillator spectrum is appropriate for the description of the new particles, their Regge slope has to be changed: $\alpha'_{new} \approx \frac{1}{2}\alpha'_{old}$ . Further it seems that the first radial excitations (r=1) are systematically lower than the second Regge recurrences. Such a deviation from the oscillator spectrum was predicted by the charmonium model [10].

The change of a fundamental dimensional parameter like the Regge slope $\alpha'$ indicates that the breaking of SU(4)-symmetry might be more complex than SU(3)-breaking. Do our dynamical relations Eq. (6) and (7) shed some light on this problem. For this we consider the leptonic decay constants in Table II. There we compare the experimental values with the theoretical values derived from Eq. (7), Z=1, with a universal and a modified Regge slope. The result is unsatisfactory. There is no indication of an $\sqrt{r+1}$ -dependence of the $g_V$ of radially excited states. The absolute value is off by a factor 3-4. On the other hand, the modification of $\alpha'$ goes in the right direction. Therefore, it seems that SU(4)-symmetry breaking is accompanied by a change of the size of the particles [11].

Table I: New Particles

| $j^{\pi c}$ | $\ell$ r | $\bar{p}p-\bar{n}n$ | $M^2$ | $\Delta M^2$ | $\bar{c}c$ | $M^2$ | $M^2$ |
|---|---|---|---|---|---|---|---|
| $0^{-+}$ | 0 0 | (140) | 0.02 | | X(2.8) | 7.84 | |
| | | | | 0.57 | | | 1.77 |
| $1^{--}$ | 0 0 | (770) | 0.59 | | J(3.1) | 9.61 | |
| | | | | 1.13 | | | 2.44 |
| $2^{++}$ | 1 0 | $A_2$(1310) | 1.72 | | $P_c$(3.41) | 11.63 | |
| | 1 0 | | | | (3.53) | 12.46 | |
| | | | | 0.84 | | | 1.64 |
| $1^{--}$ | 0 1 | (1600) | 2.56 | | (3.7) | 13.69 | |
| | | | | 1.10 | | | |
| $3^{--}$ | 2 0 | g(1680) | 2.82 | | | | |
| | | | $GeV^2$ | $GeV^2$ | | $GeV^2$ | $GeV^2$ |

$j^{\pi c}$, $\ell$, r spin parity, orbital angular momentum, and radial excitation quantum numbers in a quark model.

But we should keep in mind, that our understanding of the new particles is still very preliminary. Thus we disregarded the vector meson resonances around 4.1 GeV, because the experiments do not yet say how many resonances are there. There are now reliable data on strong couplings, which allow a comparison with dynamical relation Eq. (6). On the other hand, we have to be prepared that the new particles deeply affect our dynamical assumptions.

## 4. REVIEW OF THE BASIC DYNAMICAL ASSUMPTIONS OF THE HEAVY QUARK MODEL

New experimental and theoretical discoveries motivate a critical discussion of the assumptions involved in our phenomenological approach to a relativistic quark model.

The model is formulated in the language of general field theory. It describes the bound states with help of the B.S. amplitudes, Eq. (4), which satisfy a general B.S. equation:

Table II: Leptonic Decays

| $V$ | $\rho^0$ | J | $\psi'$ | |
|---|---|---|---|---|
| $\Gamma(V \to e^+e^-)$ | 6.6 | 4.8 | 2.2 | keV |
| exp. $g_V$ | 0.11 | 0.8 | 0.7 | $\mathrm{GeV}^2$ |
| $g_{V\,\mathrm{theor.}}$ | 0.15 | 0.1 | 0.14 | $\alpha' = 1\ \mathrm{GeV}^2$ |
| $g_{V\,\mathrm{theor.}}$ | | 0.2 | 0.3 | $\alpha' = 0.5\ \mathrm{GeV}^2$ |

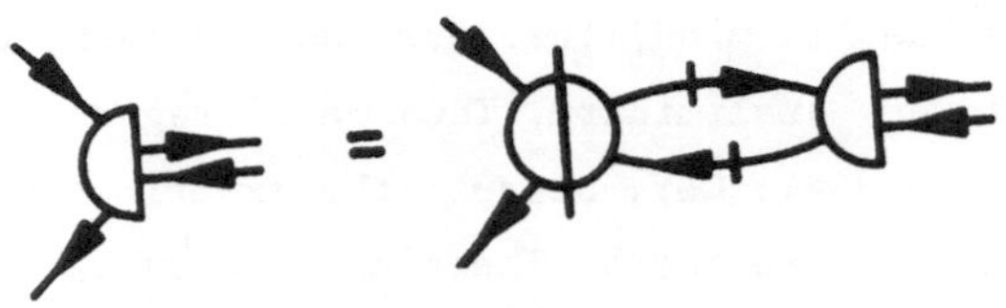

$$S^{-1}(P_1)\chi(q,P)\bar{S}^{-1}(P_2) = i\int dq' \mathfrak{K}(q,q';P)\chi(q',P) \tag{8}$$

Of course it is the hope, to derive sometimes the B.S.-kernel $\mathfrak{K}$ , or directly the B.S. amplitudes, from a Lagrangean, f.i. the "Standard Lagrangean"

$$\mathcal{L} = -\frac{1}{4} F_{\mu\nu} F^{\mu\nu} + i\bar{\Psi}\not{D}\Psi + \bar{\Psi}M\Psi$$

containing gluon fields $F_{\mu\nu}$ and quark fields $\Psi(x)$ coupled by a minimal SU(3)-colour gauge invariant interaction term. At the moment we do not have the methods for such a derivation of $\mathfrak{K}$ . Therefore we have to make assumptions on $\mathfrak{K}$ , guided by simplicity and phenomenological feed back, and the hope that an approximate $\mathfrak{K}$ guessed in this way might give a link between phenomenology and the fundamental theory.

Let me comment these assumptions:

(1) Heavy free quarks: $M_{Mes} \ll m_q \; (\sim \infty ?)$ , considered as an pragmatic model of "quark" confinement [12].

This assumption is implicitely contained also in the bag and string models [13].

(2) The inverse propagator $S^{-1}(q) = \gamma q - m$ for simplicity.

If the following expansion exists, this can be considered as an approximation appropriate in the case of large $m$ :

$$S_F'(p) = -\int_m^\infty ds \frac{\omega_1(s) + \gamma p\, \omega_2(s)}{p^2 - s + i\varepsilon}$$

$$\approx \int_m^\infty \frac{ds}{s}\,\omega_1(s) + \gamma p \int_m^\infty \frac{ds}{s}\,\omega_2(s) + p^2 \int_m^\infty \frac{ds}{s^2}\,\omega_1(s)$$

$$= \omega_{11} + \gamma p\, \omega_{21} + p^2 \omega_{12}$$

therefore

$$\left( S_F'(p) \right)^{-1} = Z\left( \not p - \hat m - \frac{1}{\varkappa} p^2 \right)$$

with $Z$, $\hat m$, $\varkappa$ expressed by the $\omega_{ik}$.

As a critical remark we want to add, that the quark 2-point function is not gauge invariant. In so far as field theoretical confinement schemes strongly relay on gauge invariance [14], this might rise the question of the physical meaning of the propagator in our phenomenological model.

(3) The kernel is of convolution type $\breve{\mathfrak{K}}(q-q')$ for simplicity.

In spite of the similarity to perturbation theoretic one particle exchange, we do not suggest that the interaction is produced by this mechanism. On the contrary, we consider $\breve{\mathfrak{K}}$ an effective kernel, originated from the collective field interactions like in the formation of bags.

Convolution type kernels cannot produce stability at bound state mass zero. Recently G. Wanders [15] has shown, that ladder graphs with strong coupling generate tachions. This is a serious argument against purely convolution type kernels. Furthermore, I believe that this assumption has to be revised in view of the PCAC problems discussed in Sect. 6.

(4) The spin dependence of $\breve{\mathfrak{K}}$.

With help of group theoretical methods, we performed an analysis of the general ansatz

$$\mathcal{K}(q) = K^P(q)\,\gamma_5 \times \gamma_5 + K^A\,\gamma_5\gamma_\mu \times \gamma_5\gamma_\mu + K^T\sigma_{\mu\nu} \times \sigma_{\mu\nu}$$
$$+ K^V\,\gamma_\mu \times \gamma_\mu + K^S\,1 \times 1 .$$

As result we propose 3 models with the following interactions

A) $-\frac{1}{4}\left[\gamma_5 \times \gamma_5 + \gamma_\mu \times \gamma_\mu - 1 \times 1\right] \cdot K(q)$

C) $\gamma_5 \times \gamma_5 \cdot K(q)$ K(q) attractive

E) $-\gamma_5 \times \gamma_5 \cdot K(q)$

which reproduce for the low lying states the structure of the meson spectrum, known from the non-relativistic quark model.

(5) Wick rotation:

Analytic continuation $q_0 \to i q_4$ is allowed for B.S. equations with kernels which are superpositions of one particle exchanges. The "Wick-rotated" equation can be more easily approximated.

(6) $O^{\pi C}(4)$-symmetry.

After Wick-rotation we make use of the assumption $m_q^2 \gg M^2$, and consider the B.S.-equation with $M^2 = 0$ as a first approximation. This equation is symmetric under 4-dimensional rotations of the relative momentum and the spins, as well as under space reflection and charge conjugation. A decomposition of the B.S. amplitudes corresponding to irreducible representations of this group $O^{\pi C}(4)$ separates the 16-component fermion-antifermion B.S. equation in hyper-radial equations [16]. There are six systems of such equations, characterized by intrinsic parities and charge parities, of which three are simple equations and three systems of three coupled equations. This group theoretical analysis allows a survey on the quantum numbers of the solutions.

(7) "Smooth potentials"

The Fourier transform of a Wick-rotated convolution type B.S.-

kernel might be considered as a 4-dimensional potential. In order that the level spacing is small compared to the quark-mass this potential must be smooth [17]. (Fig. 1). This has been shown by numerical calculations with spinless B.S.-equations [18]. For smooth potentials one has $\langle p^2 \rangle$, $\langle q^2 \rangle \ll m_q^2$. This justifies under certain conditions the approximations of the propagator S and the kernel $\tilde{\mathfrak{K}}$ mentioned in (2) and (3). Because of Ward identities, $\tilde{\mathfrak{K}}$ and S are closely related [19]; their approximations should be consistent.

(8) "Oscillator approximation"

Now we have the situation illustrated in Fig. 1. It is evident that for low-lying states the potential can be approximated by harmonic forces [20]. Since the oscillator potential "confines" the quarks [21], this approximation scheme illustrates how the limit $m_q \to \infty$ approaches a confinement scheme. In our case, the resulting wave functions are good approximations for relative spacelike momenta, but their analytic continuation into the time-region is not justified. One should expect related difficulties in the field theoretical interpretation of confinement potentials.

When we take all these considerations together, we get the dynamical equation

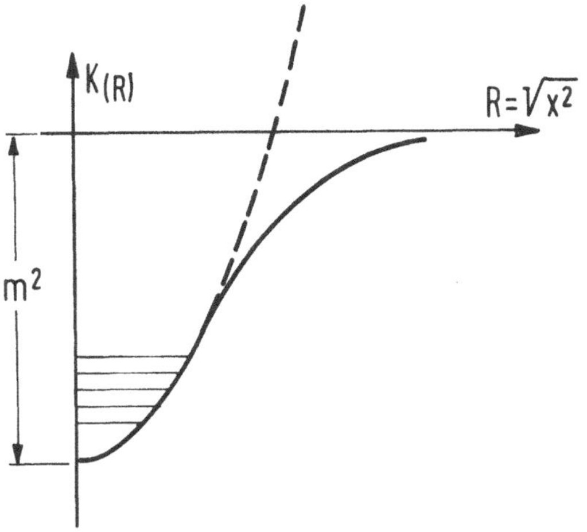


Fig. 2. Oscillator approximation of a deep smooth potential

$$(\gamma q - i m_q)\chi(\gamma q - i m_q) = (\alpha - \beta \Box_q)\gamma\chi \tag{9}$$

as an approximation to the M = 0 B.S.-equation

$$\begin{aligned}&\left(\tfrac{i}{2}M\gamma_4 + \gamma q - i m_q\right)\chi\left(-\tfrac{i}{2}M\gamma_4 + \gamma q - i m_q\right)\\ &= (\alpha - \beta\Box_q)\gamma\chi + \ldots\end{aligned} \tag{10}$$

From this we get the spectrum, already described in Eq. (1), and table of B.S.-amplitudes on which we based our phenomenological considerations of Sect. 2.

(9) Mass-compensation

The mechanism by which solutions of Eq. (9) are generated is the following: the depth of the potential $\alpha$ has to compensate $m_q^2$, the spin term $\gamma$ must be attractive for such a Dirac structure of $\chi$, for which the terms of order $\langle q\rangle m_q$ vanish. For example, in model A we have

$$\gamma\gamma_5 = \gamma_5$$

Therefore, with $\chi = \gamma_5 F(q)$ we get in (9)

$$(\gamma q - im)\gamma_5(\gamma q - im)F(q) = (\alpha - \beta\Box_q)\gamma_5 F(q)$$

or

$$-(q^2 + m^2)F(q) = (\alpha - \beta\Box_q)F(q)$$

If $\alpha = -m^2 + E$, we rewrite this again for $\chi = \gamma_5 F(q)$

$$\gamma q\,\chi\,\gamma q + \beta\Box_q\chi = E\chi$$

This might be considered as the description of the motion of a massless quark "inside" the bag generated by the confining oscillator potential. This example illustrates also how the assumption of a heavy mass determines essentially the spin structure of the B.S. amplitude.

(10) Mesonic interactions and "Saturation" of quark forces

The formulation of our relativistic quark model in the notion of general field theory enables us to make a consistent ansatz for the calculation of mesonic coupling constants, in which all mesons are treated symmetrically [22]. As a model, we may calculate these constants by evaluation of the triangle graph Fig. 1 with help of our B.S. amplitudes. The forces between mesons are then derived from the interactions between quarks. They will increase in general with the quark mass. Only one of our models (cf (4) ), namely model E, shows "saturation" of the quark forces, i.e. the mesonic couplings do not depend on the quark mass in the limit of large $m_q$. We consider this an argument in favour of the $-\gamma_5 \times \gamma_5$ interaction.

## 5. COMMENTS ON BAG MODELS

The characteristics of quark dynamics should become clearer by the comparison of different models. Let us therefore have a look on bag models and compare them with the qualitative features of the dynamics described above.

In order to have a base for such a discussion, I shall present shortly the 2-dimensional model [23], from which the SLAC-Bag is developed. The field equations of this model are

$$(-\partial_t^2 + \partial_y^2)\phi(x) + m^2\phi(x) - \lambda\phi^3(x) = -g\bar{\psi}(x)\psi(x)$$

$$i(\gamma_0\partial_t + \gamma_1\partial_y)\psi(x) + g\phi(x)\psi(x) = 0$$

$$x = (t, y) \tag{11}$$

These classical equations have a time independent solution, where the scalar field $\phi(x)$ is of "kink" form

$$\phi_{Kl}(x) = \frac{m}{\sqrt{\lambda}} \tanh \frac{my}{\sqrt{2}} \tag{12}$$

and the spinor field $\Psi(x)$ describes a quark trapped by the kink

$$\Psi_{Kl}(x) = N \left[\cosh \frac{ym}{\sqrt{2}}\right]^{-\sqrt{\frac{2}{\lambda}} \cdot y} \begin{pmatrix} 1 \\ -i \end{pmatrix}$$

$$\bar{\Psi}_{Kl} \Psi_{Kl}(x) = 0 \tag{13}$$

The Hamiltonean of these field equations shows spontaneous symmetry breaking. The kink solution connects the two stable ground state values $\phi_0 = \pm m/\sqrt{\lambda}$, $\Psi = 0$ Therefore it is topologically stable, and characterized by the topological quantum number

$$K = [\phi(+\infty) - \phi(-\infty)]\sqrt{\lambda}/2m = 1.$$

We discuss the semi-classical quantization of these fields, disregarding the quark field. For this we linearize the scalar field in the neighbourhood of the kink solution:

$$\phi(x) = \phi_{Kl}(x) + e^{i\omega_n t} \varphi_n(y) \tag{14}$$

From Eq. (11) we get the linearized equation of Schrödinger type, with a potential $\sim \phi_{Kl}^2(y)$:

$$\left(-\frac{d^2}{dy^2} - m^2 + 3\lambda \phi_{Kl}^2(y)\right) \varphi_n(y) = \omega_n^2 \varphi_n(y) \tag{15}$$

Also this equation can be solved explicitely, the eigenvalues and eigenfunctions are:

$$\omega_0^2 = 0 \qquad \varphi_0(y) = \left[\cosh \frac{my}{\sqrt{2}}\right]^2 \sim \frac{d}{dy} \phi_{Kl}(y)$$

$$\omega_1^2 = \frac{3}{4}\mu^2 \qquad \varphi_1(y) = \sinh\frac{my}{\sqrt{2}} / \varphi_0(y)$$

$$\omega_2^2 = 4\mu^2 \qquad \varphi_2(y) = \ldots \tag{16}$$

and for the continuous spectrum

$$\omega_p = \sqrt{p^2+\mu^2} \qquad \mu^2 = 2m^2$$

$$\varphi_p(y) = e^{ipy}\left[3\tanh\frac{ym}{\sqrt{2}} + \left(\frac{p}{\mu}\right)^2 - 1 - 3i\frac{p}{\mu}\tanh\frac{ym}{\sqrt{2}}\right.$$

$$\rightarrow e^{\pm i\delta} e^{ipy} \qquad \text{for } y \rightarrow \pm\infty \tag{16'}$$

These solutions have the following meaning:

$\omega_0 = 0$ represents the translation mode of the kink.

$\omega_1, \omega_2$ are classical excitation frequencies of the king.

$\omega_p$ are the frequencies of the scattering mode. It gives the correct energy momentum relation of the asymptotically free mesons.

The zero frequency translation mode is classically unstable. Therefor it must be separated from the other modes with help of a canonical transformation, as extensively discussed by Christ and Lee [24]. This transformation is rather involved, as we can imagine from the similar, but simpler problem treated here by G. Morpurgo [25]. We sketch here only the main feature. The canonical transformation, at t = 0, between the original field coordinates A(x), A(x), and the new coordinates Z("kink position"), P(total momentum), $\pi(\bar{p})$, $q(\bar{p})$ (canonical coordinates of the remaining modes) with the leading terms

$$A(y) = \phi_{Kl}(y-Z) + \int \varphi_{\bar{p}}(y-Z)\, q(\bar{p})\, d\bar{p}$$

$$\dot{A}(y) = -P\left[M_{K1}^2+P^2\right]^{-1/2} \frac{\partial \phi_{K1}(y-Z)}{\partial y}$$
$$+\int \varphi_{\bar{p}}(y-Z)\,\pi(\bar{p})\,d\bar{p} + \dots \tag{17}$$

brings the Hamiltonian in a form

$$H = \frac{1}{2}\int dy\left(\dot{A}^2+\left(\frac{d}{dy}A\right)^2 - m^2A^2+\frac{\lambda}{2}A^4\right)$$
$$= \sqrt{P^2+M_{K1}^2} + \frac{1}{2}\sum_{i=1}^{2}\left(\pi_i^2+\omega_i^2 q_i^2\right)$$
$$+\frac{1}{2}\int dp\left(\pi^2(p)+\omega^2(p)\,q^2(p)\right) \tag{18}$$

which agrees with the classical interpretation of the modes discussed above. $M_{K1} = (\sqrt{2}m)^3/3\lambda$ is the classical kink energy. The quantum mechanical kink state $|K1\rangle$ is characterized by the absence of excited modes: $2^{-1/2}[\sqrt{\omega}\,q(p) + i\pi(p)/\sqrt{\omega}]\,|K1\rangle = 0$ Implementing formally the canonical transformation (17) by a "dressing" operator, leads to a representation of $|K1\rangle$ by a coherent state:

$$|K1\rangle \sim \,:\exp\left\{-i\int dy\left(\phi_{K1,v}(x)\frac{\partial}{\partial t}A(x) - A(x)\frac{\partial}{\partial t}\phi_{K1,v}(x)\right)\right\}:|0\rangle$$

$$\phi_{K1,v}(x) = \phi_{K1}(\gamma(y-Z-vt)),\quad \gamma = 1/(1-v^2) \tag{19}$$

So much on the formal treatment of field theoretical bags. The "realistic" SLAC-bag is a 4-dimensional version of the model defined by Eq. (11). In three space dimensions the kink gets deformed into a hole in the spontaneously broken vacuum. It is no longer topologically stable, but only energetically: $E(\text{bag} + n\text{ quarks})$ is much smaller than the mass of the free quarks: $n \cdot gm/\sqrt{\lambda}$ . The other bags [6] are variations of this general pattern, or represent limits [26] of field theoretical bags (f.i. the "MIT-bag").

A comparison of the dynamics of bags with that of the heavy quark model (Sect. 4) shows similarities and rises some formal questions:

Quark confinement appears as a limiting case of a large free quark mass.

Excitations of the "bag" state are determined classically by the potential $V(y) = 3\lambda\phi^2_{K1}(y)$ in Eq. (15). The smoothness of this potential follows from the special properties of the field equations, which allow the kink solution; this means in particular the spontaneously broken symmetry.

Bags are described mainly by classical solutions of non-linear field equations. They might be represented approximately by coherent states. Their description by B.S.-amplitudes suggested by the heavy quark model, rises some problems: Since the kink solution "connects" the two stable vacua $|+\rangle, |-\rangle$ we get for all matrixelements $\langle \pm | \phi(x_1) \ldots \phi(x_n) | K1 \rangle = \mathcal{O}$. Therefore quark fields describing bag type states with help of B.S. amplitudes must be chosen carefully. Maybe $\bar{\psi}(x)\psi(x')$ in the definition, Eq. (4), of the phenomenological B.S. amplitude must be interpreted as a gauge invariant product $\bar{\psi}(x) \exp ig \int_x^{x'} A_\mu(\xi) d\xi^\mu \psi(x')$. The recent discussion [27] of solitons described by local Thirring fields should also shed some light on this problem.

Anyhow, in view of the many applications of the heavy quark model, a more deeper study of the relations between these two approaches seems to me very promising.

## 6. ON CHIRAL SYMMETRY AND PCAC

Let me mention finally a problem of our model which is related to the much discussed topic of "current and constituent" quarks. This is the question of the meaning of chiral symmetry and PCAC for the quark model of the hadron spectrum.

We consider chiral symmetry as spontaneously broken symmetry realized in nature in the limit of vanishing pion mass.[28] In our heavy quark model, the limit of zero bound state mass plays an essential rôle (cf. Sect. 4, (6) ). Therefore we could consider a model with exact chiral symmetry. But a look at our saturating model E reveals the following situation. The wave function of the pion is

$$\chi^{\pi^+}(q,P) = \frac{4\pi}{\sqrt{\beta}}\left[\gamma_5 - \frac{\not P}{M}\left(1-\frac{\not P}{2m_q}\right) - \frac{1}{2}\varepsilon^{\mu\nu\varrho\sigma}\sigma_{\mu\nu}\frac{P_\varrho q_\sigma}{M m_q}\right]e^{-q^2/2\sqrt{\beta}}\,|p\bar{n}\rangle \tag{20}$$

It is normalized according to the condition for B.S.-amplitudes

$$-\frac{1}{(2\pi)^4}\,\mathrm{Trace}\int d^4q\,\bar{\chi}_r(q,P)\left(\frac{\not P}{2}-i\not q-m_q\right)\chi_{r'}(q,P)\left(\frac{\not P}{2}+i\not q+m\right)$$

$$= \lambda\left.\frac{dP^2}{d\lambda}\right|_{P^2=M^2}\cdot\delta_{rr'} \tag{21}$$

$\lambda$ coupling constant in the kernel. This leads to a pion decay coupling constant defined as

$$\langle 0|j_\mu^{+A}(0)|{\pi^+ \atop P}\rangle = i(2\pi)^{-3/2} f_\pi P_\mu \cdot$$

$$= (2\pi)^{-1/2} Z\,\mathrm{Trace}\int \gamma_5\gamma_\mu\chi^{\pi^+}(q,P)\,d^4q .$$

which has the value

$$f_\pi = \frac{\alpha'^{-1}}{2\pi M_\pi}\cdot Z \approx (1.6\ \mathrm{GeV})\cdot Z$$

$$f_\pi^{exp} \approx 130\ \mathrm{MeV}$$

Since in the e.m. interaction we have $Z\approx 1$, there is a serious disagreement with experiment. The reason for this is an even more serious theoretical failure, namely the limit $M_\pi \to 0$ does not exist, there is no PCAC.

I would like to express some opinions to this important, unsolved problem of the quark dynamics of the meson spectrum.

In the calculation above, violation of PCAC results from the normalization condition in connection with the special spin structure of the B.S. amplitude. For example, in the model A with P + V - S interaction, the leading part of $\chi^{\pi}$ is of $\gamma_5$-type. Here we get $f_\pi = \sqrt{3}/2\pi\, m_q \alpha'$ PCAC is not violated, and for $m_q$ = 3 GeV we even would reproduce the experimental value. But model A is not saturating, and therefore does not lead to reasonable values for hadronic decays.

The normalization condition is determined on the right hand side by $dP^2/d\lambda = c \approx \alpha'^{-1}$ For a model, stable at bound state mass zero (cf. Sect. 4, (3) ), we would expect $c \to o$ for $p^2 \to o$. This would change $f_\pi$ in the right direction. Therefore our PCAC-problem seems to be related to the instability problem of convolution type kernels.

There is the obvious argument, that in a theory with large quark masses, chiral symmetry will have some difficulties. But as already mentioned, the mass we are talking about, is not a mass appearing in the Lagrangean, but the mass of a free quark (cf. Sect. 4, (9) ), which might be generated by spontaneous symmetry break down. The question we discuss here is the problem of the conciliation of chiral symmetry with quark confinement.

REFERENCES

1) M. Böhm, H. Joos and M. Krammer,
a) Nuovo Cimento 7 A, 21 (1972)
b) Nuclear Phys. B51, 397 (1973)
c) Acta Phys. Austriaca 38, 123 (1973)
d) in Recent developments in mathematical physics (P. Urban, ed.) (Springer Verlag, Wien and New York, 1973), pp. 3-116
e) Nuclear Phys. B69, 349 (1974)
f) Report DESY 74/7 (1974)
M. Böhm and M. Krammer,
g) Phys. Letters 50 B, 457 (1974).

2) H.-J. Lipkin, Physics Reports 8c, 3 (1973)

3) M. Böhm, H. Joos, M. Krammer, CERN-Th 1949(1974)

4) J.J. Aubert et al., Phys. Rev. Lett. 33 (1974) 1404
J.E. Augustin et al., Phys. Rev. Lett. 33 (1974) 1406
C. Bacci et al., Phys. Rev. Lett. 33 (1974) 1408
W. Braunschweig et al., Physics Lett. 55B (1974) 393
L. Criegee et al., Physics Lett. 55B (1974) 489

5) New results on the baryons in the relativistic heavy quark model will be reported by M. Böhm and R. Meyer at this meeting

6) a) A. Chodos, R.L. Jaffe, K. Johnson, C.B. Thorn and V.F. Weisskopf, Phys. Rev. D9 (1974) 3471. (MIT-bag)
b) W.A. Bardeen, M.S. Chanowitz, S.D. Drell, M. Weinstein, T.M.Yan, Phys. Rev. D11 (1975) 1094
c) P. Vinciarelly; Nuov. Cim. Lett. 4 (1972) 905, Nucl. Phys. B89 (1975) 463

7) Particle Data Group; Rev. Mod. Phys. 47, 535 (1975)

8) P. Langacker, G. Segre; University of Pennsylvania preprint UPR-0049T (1975)
F. Ceradini et al., Phys. Lett. 43 B, (1973) 341

9) Ref. 4
DASP-Collaboration, Physics Lett. 57B (1975) 407
SLAC-LBL-Group, Phys. Rev. Lett. 35 (1975) 821
J. Heintze, DESY 75/34
B.H. Wiik, DESY 75/37

10) T. Appelquist, A. De Rújula, H.D. Politzer, S.L. Glashow, Phys. Rev. Lett. 34 (1975) 365
E. Eichten, K. Gottfried, T. Kinoshita, J. Kogut, K.D. Lane, T.M. Yan, Phys. Rev. Lett. 34 (1975) 369

11) M. Böhm, M. Krammer, DESY 74/52
M. Krammer, H. Krasemann, DESY 75/19
H. Krasemann, Diplomarbeit Hamburg 1975
P. Becher, M. Böhm, Preprint Univ. Würzburg

12) Practical quark confinement by large quark masses was introduced in a non-relativistic framework by G. Morpurgo, Physics 2, 95 (1965). This viewpoint was recently stressed in the context of field theory by G. Preparata, Phys. Rev. D7, 2973 (1973), and by M. Böhm, H. Joos and M. Krammer, particularly in Ref. 4d

13) A. Casher, J. Kogut, L. Susskind, Phys. Rev. Lett. 31 (1973) 792 and J.H. Lowenstein, J.A. Swieca, Ann. Phys. 68 (1971) 172
W.A. Bardeen et al., Ref. 6b)

14) K. Wilson, Phys. Rev. D10 (1974) 2445
A. Casher et al., Ref. 13

15) G. Wanders, Physics Letters 58B (1975) 191

16) S. Mandelstam, Proc. Roy. Soc. A237, 496 (1956)

17) M.K. Sundaresan and P.J.S. Watson, Ann. Phys. (N.Y.) 59, 375 (1970)
G. Preparata, Subnuclear phenomena (Academic Press, New York and London, 1970), p. 240

18) A. Pagnamenta, Nuovo Cimento 53 A, 30 (1968). See also Ref. 4a.
P. Narayanaswamy and A. Pagnamenta, Nuovo Cimento 53 A, 635 (1968). See also Ref. 4a.

19) M. Böhm, CERN Th. 1878

20) Relativistic dynamical models with harmonic forces are discussed for instance by R.P. Feynman, M. Kislinger and F. Ravndal, Phys. Rev. D3, 2706 (1971).
We would like to stress that we consider harmonic forces as approximations in the Wick-rotated BS equation

21) K. Johnson, Phys. Rev. D6 (1972) 1102

22) Compare R.P. Feynman et al., Ref. 20

23) R. Dashen, B. Hasslacher, A. Neveu, Phys. Rev. 10D (1974) 4130

24) N.H. Crist, T.D. Lee, Columbia Univ. New York, CO 2271-55

25) G. Morpurgo, Erice Lectures 1974

26) M. Creutz, Phys. Rev. D10 (1974) 1749

27) S. Coleman, Phys. Rev. D11 (1975) 2088
S. Mandelstam, Phys. Rev. D11 (1975) 3026

28) S. Weinberg-Proceedings of the 14th International Conference on High Energy Physics, Vienna (1968) p. 253

# RELATIVISTIC QUARK EQUATIONS, INSTANTANEOUS INTERACTIONS AND QUASI-POTENTIALS

R.G. Moorhouse
Glasgow University, Scotland

There are at least two reasons why one would wish to investigate relativistic equations in constituent quark model hadron spectroscopy calculations. The first is that in calculating matrix elements for processes such as $N^* \rightarrow N + \gamma$ at least one of the hadrons must be boosted and so we require either a Lorentz covariant description or, at least, a prescription for a boost. This problem shows itself in simplest form in the calculation of nuclear form factors and here 2nd generation quark models of the 4-dimensional oscillator type such as that of Feynman, Kislinger and Ravndal[1] have so far failed. What I have to say is not directly relevant to these problems. The second reason is that in constituent quark model calculations, with non-charged quarks, the effective quark mass is small[2,3] of the order of 300 or 400 MeV which is the order of the internal momenta of the quarks in such calculations. There is thus a need for at least the investigation of relativistic equations, involving such a quark mass, in the first place for mass spectrum calculations, this talk is concerned with that problem.

If one adopts a manifestly covariant approach, such as the Bethe-Salpeter equation with a manifestly covariant interaction kernel (potential) unobserved time-like excitations may occur[4]; and it may not be so easy to make the connection with the Schrödinger, or other 3-dimensional equations while one might well wish to make this connection in view of the phenomenological first-order success of 3-dimensional oscillator equations and the fact that our experience commonly lies in the more tractable 3-dimensional domain. This indicated to us an approach by the Salpeter equation[5], which is the 3-dimensional form to which the Bethe-Salpeter equation reduces when the interaction kernel represents an instantaneous interaction, or an approach by the formally similar, but motivationally different, quasi-potential equations typified by the Logunov-Tavkhelidze-Blankenbecler-Sugar equation[6,7].

## THE SALPETER EQUATION FOR MESONS

The homogeneous Bethe-Salpeter equation for two equal mass spin-½ particles, a and b, of relativistic momentum $q = \frac{1}{2}(p_a - p_b)$ and total 4-momentum $P = p_a + p_b$ is

$$(\not{p}_a - m)(\not{p}_b - m)\Psi(q) = -\int V(q,q')\Psi(q')d^3q'dq_0' \,. \tag{1}$$

An instantaneous potential V is in momentum space independent of the relative energies $q_0$, $q_0'$ so defining

$$\Phi(\underline{q}) \equiv \frac{1}{2\pi i}\int dq_0\Psi(\underline{q},q_0) \tag{2}$$

We derive from (1) the equation

$$\Phi = -\frac{1}{2\pi i}\int dq_0 \times$$

$$\times\frac{\left(\frac{1}{2}\cdot E + q_0) + \underline{\alpha}^a\cdot\underline{p}_a + \beta^a m\right)\left(\frac{1}{2}E - q_0 + \underline{\alpha}^b\cdot p_a + \beta^b m\right)\beta^a\beta^b V(\underline{q},\underline{q}')\Phi(q')}{\left[\left(\frac{1}{2}E + q_0\right)^2 - (p_a^2 + m^2)\right]\left[\left(\frac{1}{2}E - q_0\right)^2 - (p_b^2 + m^2)\right]} \tag{3}$$

where $E = P_0$. Performing the $p_0$ interaction and writing the result in the centre-of-mass system so that $\underline{p}_a = -\underline{p}_b = \underline{q}$ and $\sqrt{p_a^2 + m^2} = \sqrt{p_b^2 + m^2}$ or $W_a = W_b \equiv W$ we obtain

$$\Phi(q) = \int\left[\frac{\Lambda_+^a\Lambda_+^b\beta^a\beta^b V(q,q')}{(E - 2W)} - \frac{\Lambda_-^a\Lambda_-^b\beta^a\beta^b V(q,q')}{(E + 2W)}\right]\Phi(\underline{q}')d^3q' \tag{4}$$

or equivalently

$$(E - H_a - H_b)\Phi(\underline{q}) = \int[\Lambda_+^a\Lambda_+^b - \Lambda_-^a\Lambda_-^b]\beta^a\beta^b V(q,q')\Phi(\underline{q}')d^3q' \tag{5}$$

where $H_a = \underline{\alpha}^a\cdot\underline{p}_a + \beta^a m$, $H_b = \underline{\alpha}^b\cdot p_b + \beta^b m$ and the $\Lambda$'s are the usual projection operators for positive and negative energy spinors: $\Lambda_\pm = \frac{1}{2}(1 \pm H/W)$

In equation (4) or (5) E is the eigenvalue that we wish to find, being the mass of the composite system. The equations are deduced using an instantaneous potential and of course a potential instantaneous in all Lorentz frames is not a relativistic concept; the deduction of Eqs. (4) and (5) assumes that the potential is instantaneous (presumably approximately) in the centre-of-mass system. Such awkward questions are avoided, at any rate in their most direct form, in the Logunov-Tavtchelidze-Blankenbecler-Sugar[6,7] approach outlined next.

## QUASI-POTENTIAL EQUATION FOR MESONS

Faustov[8], following Logunov and Tavkhelidze, has derived a quasi-potential equation for spin-½ particles for application to the hyperfine structure of hydrogen. We follow his derivation, using for brevity the usual symbolic notation.

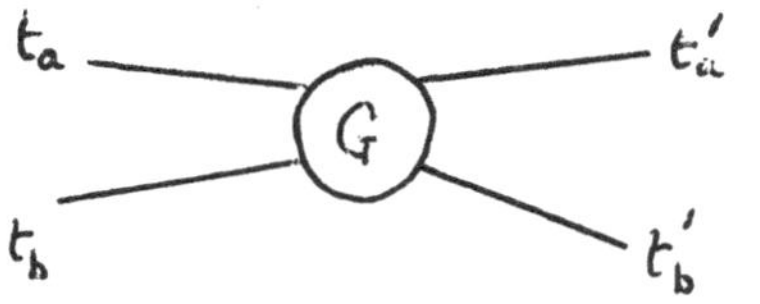


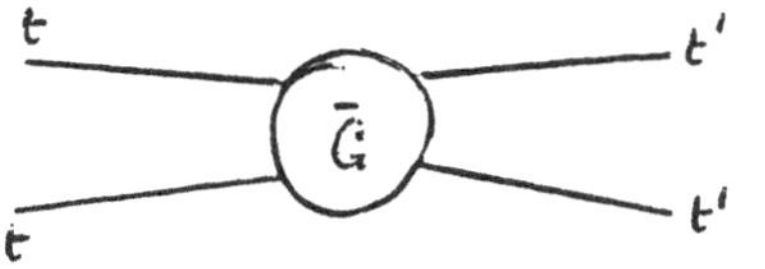


If G is the two particle Green's function and $G_0$ is the product of the 2 single particle Greens functions then let $\overline{G}$, $\overline{G}_0$ be the corresponding "two-time" Green's functions $t_a = t_b$, $t'_a = t'_b$, in momentum space

$$G = \int_{-\infty}^{\infty} d\varepsilon d\varepsilon' G(\varepsilon,\varepsilon') \qquad (6)$$

where $\varepsilon, \varepsilon'$ are related to the initial and final relative energies.

The off-mass shell two particle scattering amplitude T is given by

$$G_0 T G_0 = G - G_0 \qquad (7)$$

and we can define a similar amplitude $\overline{T}$ by

$$\overline{G}_0 \overline{T G}_0 = \overline{G} - \overline{G}_0 \ , \qquad (8)$$

$\overline{T}$ corresponds to equal times for the two initial state particles and equal times for the two final state particles; and in energy-momentum space $\overline{T}$ on the energy shell coincides with T on the mass shell, so $\overline{T}$ contains the physical scattering amplitude[8].

One may rewrite the defining equation (8) for $\overline{T}$ in the form of a quasi-potential equation

$$\bar{T} = V + V\bar{G}_0\bar{T} \tag{9}$$

where the quasi-potential V can be formally defined by

$$-V = \bar{G}^{-1} - \bar{G}_0^{-1} \ . \tag{10}$$

In full the equation (9) is

$$\bar{T}(\underline{q},\underline{p},E) = V(\underline{q},\underline{p},E) + \int V(\underline{q},\underline{k},E)F(k)\bar{T}(\underline{k},\underline{p},E)\ d^3k \tag{11}$$

where

$$\bar{G}_0(k,k') = F(\underline{k})\delta(\underline{k} - \underline{k}')$$

and we can define a wave function corresponding to $\bar{T}$ by

$$\psi(\underline{q}) = \delta(\underline{q} - \underline{p})u_a(\underline{p})u_b(-\underline{p}) + F(\underline{q})\bar{T}(\underline{q},\underline{p})u_a(\underline{p})u_b(-\underline{p}) \tag{12}$$

which, on substituting $\bar{T}$ from (11 gives rise to the wave equation for scattering:

$$\psi(\underline{q}) = \delta(\underline{q} - \underline{p})u_a(\underline{p})u_b(-\underline{p}) + F(\underline{q}) \int V(q,q',E)\psi(q') \ . \tag{13}$$

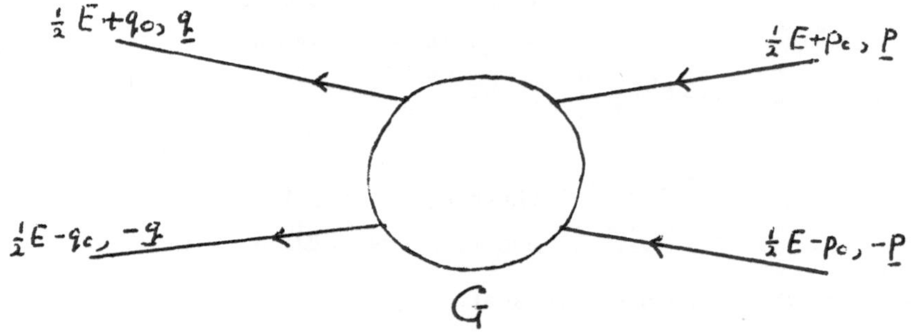


We now remember the definition of $F(\underline{q})$ in terms of the Green's functions.

$$\delta(\underline{p} - \underline{q})F(\underline{q}) = \bar{G}_0 = \int dp_0 dq_0 [G_0] = \int [G_a G_b \delta(\underline{p} - \underline{q})\delta(p_0 - q_0)]dp_0 dq_0 \tag{14}$$

where $G_a$ and $G_b$ are the <u>full</u> single particle Green's functions. If we <u>approximate</u> these by the <u>bare</u> fermion propagators then $G_a \sim 1/(\not{p}_a - m)$, $G_b \sim 1/(\not{p}_b - m)$ and, from (14),

$$F \sim F_{bare} = \left[\frac{\Lambda_+^a \Lambda_+^b}{E - W_a - W_b} - \frac{\Lambda_-^a \Lambda_-^b}{E + W_a + W_b}\right]\beta^a \beta^b \tag{15}$$

where $W_a = W_b = W$ in the centre-of-mass system. With this approximation the corresponding bound state approximation to (13) -- just ommitting the non-homogeneous plane wave part -- is

$$\psi(\underline{q}) = \left[\frac{\Lambda_+^a\Lambda_+^b}{E - 2W} - \frac{\Lambda_-^a\Lambda_-^b}{E + 2W}\right] \beta^a\beta^b \int V(\underline{q},\underline{q}',E)\psi(q')\, d^3q' \tag{16}$$

which is formally the same as Eq. (4), the Salpeter equation. But in the derivation of (16) no assumption is made that the interaction is instantaneous -- the potential V is a pseudo-potential which reproduces the two-time (equal-time) 2 particle Green's function.

However (16) is not the equation used but a further step is made. Indeed (16) is deemed inappropriate on the grounds that $\det(F_{bare}) = 0$ and so putting $\overline{G}_0 \sim F_{bare}$ in Eq. (10) which formally defines V gives rise to an undefined equation.

The further step is to define, using (12), the "positive energy spinor" part of $\psi$:

$$\psi_+(q) \equiv \Lambda_+^a(\underline{q})\Lambda_+^b(-\underline{q})\psi(q) \tag{17}$$

so that $\psi_+$ can be expressed in terms of Pauli spinors. We find from (12) that, an approximating F by $F^{bare}$ [Eq. (15)]

$$\psi_+(q) \equiv \delta(\underline{q} - \underline{p})\Lambda_+^a\Lambda_+^b u_a(\underline{p})u_b(-\underline{p}) + F_+^{bare}\overline{t}(q,\underline{p})u_a(\underline{p})u_b(-\underline{p}) \tag{18}$$

where

$$t(\underline{q},\underline{p}) = \Lambda_+^a(\underline{q})\Lambda_+^b(-\underline{q})\beta^a\beta^b\overline{T}(\underline{q},\underline{p}) \tag{19}$$

and

$$F_+^{bare}(\underline{q}) = \frac{\Lambda_+^a(\underline{q})\Lambda_+^b(-\underline{q})}{(E - 2W)} \,. \tag{20}$$

We can now find the equation for $\psi_+$ in terms of a further quasi-potential v defined by

$$v = t(1 + F_+t)^{-1} \tag{21}$$

so that

$$t = v + vF_+t \tag{22}$$

(It may be noted that t is a matrix defined between positive energy spinors on the left and positive and negative energy spinors on the right; v between positive energy spinors only on both left and right). Putting (22) in (18) gives

$$\psi_+(\underline{q}) = \delta(\underline{q} - \underline{p})\Lambda_+^a\Lambda_+^b u_a(\underline{p})u_b(-\underline{p}) + F_+(\underline{q}) \int v(q,q',E)\psi_+(q')\, d^3q' \quad . \tag{23}$$

All quantities $\psi$ and $\nu$ are now expressed in terms of Pauli spinors (positive energy spinors) all explicit reference to negative energy spinors having been deleted; these states are subsuned in the definition of v through Eq. (21). The corresponding bound state equation is

$$(E - 2W)\psi_+(\underline{q}) = \int v(\underline{q},q',E)\psi_+(q')\, d^3q' \tag{24}$$

where

$$W = \sqrt{q^2 + m^2}$$

For $q^2 << m^2$ (24) reduces to the corresponding Schrödinger equation

$$\left(E - 2m - \frac{q^2}{m}\right) \psi_+(q) = \int v(\underline{q},\underline{q}')\psi_+(q')\, d^3q' \tag{25}$$

In applications, for example, to the hydrogen atom the potential v is obtained from $\bar{T}$ via Eqs. (19) and (21), and using the S-matrix to find $\bar{T}$. In such cases, of course, simplifying approximations can validly be made.

A similar reduction of the Salpeter equation to the Pauli spinor form could be made; the question is whether the potential is of a more simple form in the full Eqs. (15), (16), or in the reduced equation (25). We have been investigating both forms with same simple potentials.

## WAVE FUNCTIONS FOR $0^-$ and $1^-$ MESONS

There is a well-known way of writing 2-body wave functions as a superposition of Dirac $\gamma$-matrices; this representation is particularly natural when the 2 bodies are particle and anti-particle and the representation of spinor outer products $u(1)\bar{u}'(2)$ or $u(2)\bar{u}'(1)$ as Dirac $\gamma$-matrices is almost immediate.

The wave function $\Phi(\underline{q})$ is a function of the relative 2-momentum $\underline{q} = \underline{p}_a - \underline{p}_b$, the relative energy having been integrated out either in the Salpeter equation or the quasi-potential apporach[9]. We consider the wave function in the centre-of-mass system.

i) For $J^{PC} = 0^{-+}$ -- that is mesons of the pseudoscalar octet.

$$\Phi(\underline{q}) = \phi_+(q^2)\gamma_5 + \phi_-(q^2)\gamma_5\beta + \phi'(q^2)\gamma_5\underline{\alpha}\cdot\underline{q} \; . \tag{26}$$

The Salpeter, or full, equation may be written

$$\Phi(\underline{q}) = \int d^3q' \left[\frac{\Lambda_+(\underline{q})\beta V\Phi(\underline{q}')\beta\Lambda_-(-\underline{q})}{E - 2W_q} - \frac{\Lambda_-(\underline{q})\beta V\Phi(\underline{q}')\beta\Lambda_+(-\underline{q})}{E + 2W_q}\right] \tag{27}$$

where

$$\Lambda_+(\underline{q}) = \frac{1}{2}\left(1 + \frac{\underline{\alpha}\cdot\underline{q} + \beta_m}{W}\right), \quad \Lambda_-(q) = \frac{1}{2}\left(1 - \frac{\underline{\alpha}\cdot\underline{q} + \beta_m}{W}\right) \quad W = \sqrt{q^2 + m^2} \; . \tag{28}$$

The momentum $\underline{q}$ is associated with particle a and $-\underline{q}$ with particle b. Since

$$\Lambda_-(\underline{q})\Lambda_+(\underline{q}) = \Lambda_+(q)\cdot\Lambda_-(\underline{q}) = 0$$

it follows that

$$\Lambda_+(\underline{q})\Phi\Lambda_+(-\underline{q}) = \Lambda_-(\underline{q})\Phi\Lambda_+(-\underline{q}) = 0 \; . \tag{29}$$

These are dynamical conditions which follow from the Salpeter equation and give

$$\phi' = -\frac{1}{m}\,\phi_- \tag{30}$$

so that we are left with

$$\Phi(\underline{q}) = \gamma_5\left[\phi + (q^2) + \phi_-(q^2)(\beta - \frac{1}{m}\,\underline{\alpha}\cdot\underline{q})\right] \tag{31}$$

and the equations of motion in two independent functions $\phi_+$ and $\phi_-$; the form of these equations depends on the $\gamma$-matrix dependence -- that is the spin-dependence-- of V. For example V could be of the form $V = (\gamma_\mu)^a(\gamma^\mu)^b\tilde{V}$ or $V = (1)^a(1)^b\tilde{V}$ where $\tilde{V}$ is $\gamma$-matrix independent. The latter example is a scalar potential and gives the equations:

$$(E^2 - 4W^2)\phi_+(q^2) =$$

$$-2W\int \tilde{V}(\underline{q},\underline{q}')\phi_-(q'^2)\; d^3q' - \frac{E}{mW}\int (m^2 - \underline{q}\cdot\underline{q}')\tilde{V}(\underline{q},\underline{q}')\phi_+(q'^2)$$

$$(E^2 - 4W^2)\phi_-(q^2) =$$

$$-\frac{2}{W}\int (m^2 - \underline{q}\cdot\underline{q}')\tilde{V}(\underline{q},\underline{q}')\phi_-(q'^2)\, d^3q' - \frac{Em}{W}\int \tilde{V}(\underline{q},q')\phi_+(q'^2)\, d^3q' \tag{32}$$

The reduced equation is, instead of (27),

$$\Phi(q) = \int d^3q' \frac{\Lambda_+^a(\underline{q})\beta V(\underline{q},\underline{q}')\Phi(q')\beta\Lambda_-^b(-\underline{q})}{E - 2W_q} \tag{33}$$

and the form of this equation imposes a further dynamical restriction on the form of $\Phi$ [Eq. (29)], since now besides (29) we have additionally

$$\Lambda_-(q)\Phi\Lambda_+(-q) = 0 \tag{34}$$

giving

$$\phi_+ = W\phi' \tag{35}$$

as well as

$$\phi' = -\frac{1}{m}\phi_- \tag{36}$$

so that

$$\Phi = \gamma_5(W + \underline{\alpha}\cdot\underline{q} - \beta m)\phi' = 2W\gamma_5\Lambda_-(-\underline{q}) .$$

Now for a scalar V the one-variable equation is

$$(E - 2W)\phi'(\underline{q}^2) = -\frac{1}{2}\int \left(1 - \frac{\underline{q}\cdot\underline{q}'}{WW'} + \frac{m^2}{WW'}\right) \tilde{V}(\underline{q},\underline{q}')\phi'(q'^2)\, d^3q' \tag{37}$$

where, as before,

$$W = \sqrt{m^2 + q^2} , \quad W' = \sqrt{m^2 + q'^2} .$$

Corresponding equations to (32) and (36) can readily be written down for various forms of $\gamma$-matrix dependence of V. For example for

$$V = \gamma_\mu^a \tilde{V}(\underline{q},\underline{q}')\gamma^{\mu b} \tag{38}$$

where $\tilde{V}$ is a scalar quantity, then instead of (37) we have

$$(E - 2W)\phi'(q^2) = \int \left(2 - \frac{m^2}{WW'}\right) \tilde{V}(q,q')\phi'(q'^2)\, d^3q' . \tag{39}$$

$1^{-}$ mesons ($\rho$, $\phi$, $\psi$, $\psi$, etc.) have more independent scalar functions, the general form in the centre-of-mass of the meson being

$$\psi_i(\underline{q}) = -\beta\alpha_i[\phi_1(q^2) + \beta\phi_2(q^2) + \underline{\alpha}\cdot\underline{q}\phi_3(q^2)] + \\ + q_i[\beta\phi_3(q^2) + \phi_4(q^2) + \underline{\alpha}\cdot\underline{q}\phi_5(q^2) + \beta\underline{\alpha}\cdot\underline{q}\phi_6(q^2)] \quad (40)$$

$i = 1, 2, 3$ being the vector index of the vector meson. The Salpeter equation imposes the dynamical conditions

$$\phi_2 = -m\phi_3 \ , \quad \phi_1 = -m\phi_4 + q^2\phi_6 \quad (41)$$

and the reduced, or quasi-potential equation, the additional conditions

$$\phi_2 = -\frac{m}{W}\phi_1 \ , \quad \phi_5 = \frac{1}{W}\phi_4 + \frac{m}{W}\phi_6 \ , \quad (42)$$

leading, for the Salpeter equation, to 4 coupled simultaneous equations in 4 independent functions and for the reduced (or quasi-potential) equation, to 2 coupled simultaneous equations in 2 independent quantities.

The $1^{--}$ wave function (40) even with the conditions (41) and (42) contains both S and D waves, the proportion of these being determined by the potential V through the equations. We can similarly write down the wave functions and wave equations which arise from orbital P-wave excitation giving rise to $0^{++}$, $1^{++}$, and $2^{++}$ mesons. For example the $0^{++}$ wave function for the Salpeter equation -- corresponding to (31) is

$$\Phi^P(\underline{q}) = (q^2 + m\beta\underline{\alpha}\cdot q)\phi_0 + \underline{\alpha}\cdot\underline{q}\phi_1 \quad (43)$$

## POTENTIALS AND SOLUTION METHODS

We can bear in mind, as a simple example, the reduced (or quasi-potential) equation for pseudoscalar mesons which is an equation in one independent function only. For example, when the potential V is $\gamma$-matrix independent we have Eq. (37):

$$(E - 2\sqrt{m^2 + q^2})\phi'(q^2) = -\frac{1}{2}\int\left[1 - \frac{(\underline{q}\cdot\underline{q}' - m^2)}{\sqrt{(m^2 + q^2)(m^2 + q'^2)}}\right]V(\underline{q},\underline{q}')\phi'(q'^2)\, d^3q' \quad (37)$$

and, of course, for a suitable V this can reduce to a non-relativistic configuration when $q^2, q'^2 \ll m^2$. In this case

$$\left[(E - 2m) - \frac{q^2}{m}\right]\phi'(q^2) = -\frac{1}{2}\int (1 - \ldots)V(\underline{q},\underline{q}')\phi'(q'^2)\, d^3q' \tag{44}$$

which is a wave equation of Schrödinger type. However (44) is not the wave equation we wish to solve; we are interested not only in charmonium type situations where $m \sim 2.0$ GeV/$c^2$ where Schrödinger equations such as (44) are valid, but also, and particularly, in the ordinary mesons of the pseudoscalar and vector nonets where the effective mass of the quarks is $m \sim 0.3$ GeV/$c^2$ and where (44) is not evidently valid. We must solve (37) and the similar more complicated equations for $1^-$, and this poses some problems. It is desirable for comparison to retain the connection with the Schrödinger (non-relativistic) situation so we consider potentials of the form $V(\underline{q}-\underline{q}')$ whose Fourier transform is an ordinary potential in configuration space

$$V(\underline{q} - \underline{q}') \to V(\underline{r})$$

where $\underline{r}$ is the inter-quark position vector. This corresponds to a Schrödinger Eq. (44) in configuration space-like

$$\left[(E - 2m) + \frac{\nabla^2}{m}\right]\phi'(r) = -\frac{1}{2}(1 - \ldots)V(\underline{r})\phi'(r) \tag{45}$$

However, we wish to solve (37) which has a different propagator $(E - 2\sqrt{m^2 + q^2})$ instead of $(E - 2m - q^2/m)$. Suppose now that V is a linear 'confining' potential

$$-V(\underline{r}) = C + kr \tag{46}$$

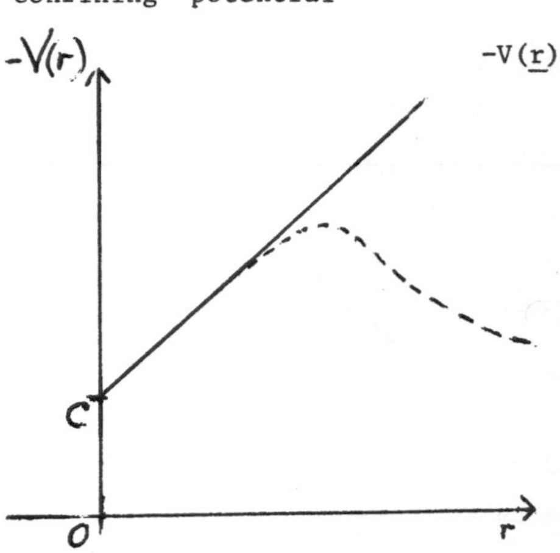


then it turns out that though the Schrödinger equation develops discrete bound states, equations of the type (37) and the other more complicated ones such as (33) that we wish to solve, only develop a continuum of states. This was a problem previously noted by Plesset[10)] in studying solutions of the Dirac equation, and applies to all polynomial type potentials. It is probably associated with negative energy states, or pairs, which are intrinsic to spin-½ particles and are contained in our relativistic treatment but not in the Schrödinger equation. Our simple solution was to use potentials of the form

$$-V(r) = C + kr\, e^{-\mu r}$$

which are approximately linear for $r < \mu$ but fall off at infinity. Physical reasons for such a screening of the potential at large distances have been given by Kogut and Susskind[11].

Another difficulty associated with large q is that the kernel $K(q,q')$ of the integral equations (37) or like the more complicated equations does not satisfy the Fredholm condition on the boundedness of $K(q,q)$. This can be corrected by making a small change in the potential and putting

$$-V(r) = C + \lambda(r\, e^{-\mu r} - \zeta^2 r\, e^{-\zeta\mu r}) , \qquad \zeta \ll 1 \tag{47}$$

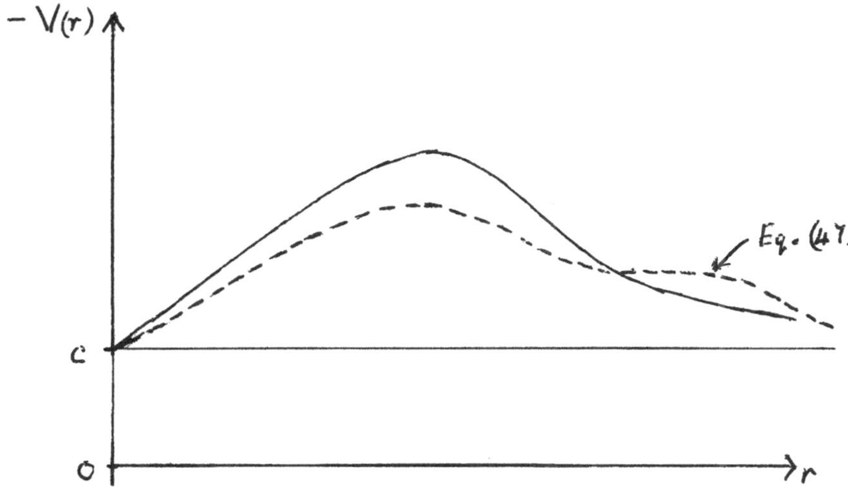


At this stage we can perform analytically the angular integration in (3.7) so that it takes the form

$$\left(E - 2\sqrt{m^2 + q^2} - \frac{Cm^2}{m^2 + q^2}\right)\phi(q^2) = \lambda \int_0^\infty A(q,q')\phi(q'^2)\, dq'$$

or

$$\phi(q^2) = \lambda \int_0^\infty B(q,q')\phi(q'^2)\, dq' \tag{48}$$

where

$$B = A \Bigg/ \left(E - 2\sqrt{m^2 + q^2} - \frac{Cm^2}{m^2 + q^2}\right) .$$

For a given E this is an eigenvalue equation in $\lambda$, the strength of the linear potential, and on taking a grid of discrete points in momentum space, the eigenvalues and eigenfunctions of equation (48) can be computed by standard matrix methods. We have also carried out a similar process and solved equations for the full simultaneous equations (32) and the simultaneous equations for the $1^-$ mesons.

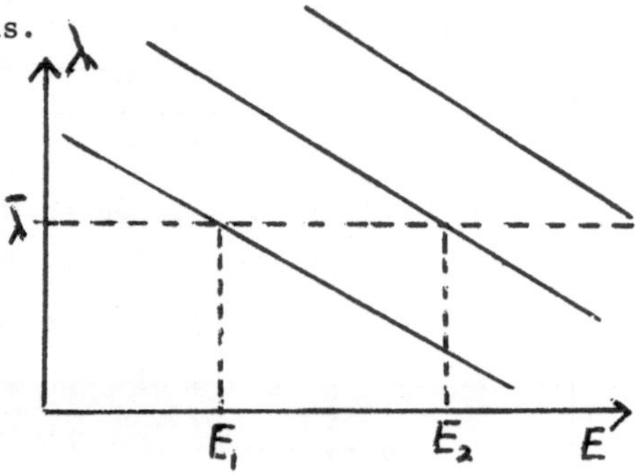


One can plot the eigenvalues $\lambda$ as functions of E, typically obtaining graphs such as the one opposite. For a given strength of potential, say $\lambda = \bar{\lambda}$ one can then read off the bound state eigenvalues of energy $E = E_1$, $E_2$, ... .

APPLICATIONS

The reduced equation (33)

$$(E - 2W_q)\Phi(q) = \int d^3q' \Lambda_+^a(\underline{q})\beta V(\underline{q},\underline{q}')\Phi(q')\beta\Lambda_-^b(-\underline{q}) \tag{33}$$

could be expanded in powers of $(v/c)^2$, and this expansion to first-order, valid in a non-relativistic situation gives the well-known relativistic corrections to the Schrödinger equation. In the case where V is Coulombic these give Fermi-Breit type Hamiltonians used, for example, by De Rújula Georgi and Glashow[3] to obtain successful relations even for uncharmed hadrons for which presumably the expansion in powers of (v/c) is not quickly covergent. Such expansions have also been used by Schnitzer[12] in the charmonium case with a potential of the form $V(r)\ k\gamma_p^a\gamma^{\mu b}r$.

On the other hand the full equation

$$\Phi(q) = \int d^3q' \left[\frac{\Lambda_+\beta V\Phi(q')\beta\Lambda_-^b}{E - 2W_q} - \frac{\Lambda_-\beta V\Phi(q')\beta\Lambda_+}{E + 2W_q}\right] \tag{27}$$

has been rather little investigated to my knowledge. In the quantum electrodynamic (positronium) applications in Schwingers "Particles, Sources and Fields", for example, the basic treatment is first to omit the last term of (27) -- that is to deal with the reduced equation (33) -- and to add in the last perturbatively as perturbation, if necessary. In certain circumstances only the last term of (27) is indeed negligable -- as for example, when V is a scalar potential and the mass of the particle is reasonably large; this latter condition being fulfilled in the charmonium configuration.

If one is prepared to consider heuristically these equations and their results for different types of potentials, there is an obvious programme of investigation: for example, one can consider (33) for the reduced or quasi-potential equation, for those potentials already considered by other authors and assess the deviations from Fermi-Breit-type Hamiltonian results in the small mass case; one can also consider the results which arise from the same potentials in using (27). Of course we do not at this moment know *à priori* whether it is appropriate to use either of these equations for hadrons, and in particular we do not know whether the full equation (27) is better than the reduced equation (33) for simple potentials. We are currently carrying out such an investigation; the work described here has been developed in collaboration with A.B. Henriques and B.H. Kellet.

## REFERENCES

1) R.P. Feynman, M. Kislinger and F. Ravndal, Phys. Rev. D3, 2706 (1971).

2) D. Faiman and A.W. Hendry, Phys. Rev. 173, 1720 (1968).

3) A. De Rújula, H. Georgi and S.L. Glashow, Hadron masses in a Gauge Theory, Harvard preprint, 1975.

4) In the weak-binding case, and for certain potentials, these unwanted excitations do not occur. For some discussion and references see N. Nakanashi, Suppl. Prog. Theor. Phys. 43, 1 (1969) pages 41-52. We are indebeted to P. Menotti for this remark.

5) E.E. Salpeter, Phys. Rev. 87, 328 (1952).

6) A.A. Logunov and A.N. Tavkhelidze, Nuovo Cimento 29, 38d (1963).

7) R. Blankenbecler and R. Sugar, Phys. Rev. 142, 1051 (1966).

8) R.N. Fausov, Nucl. Phys. 75, 669 (1966).

9) The analogous 4-dimensional forms (from which the 3-dimensional forms can be derived) have been written down and discussed by C.H. Llewellyn Smith, Ann. Phys. (New York) 53, 521 (1969) and by G. Morpurgo, Proc. of the 9th Course of the Int. School of Sub-

nuclear Physics Ettore Majorana: Properties of the Fundamental Interactions, Erice, 1971) (ed. A. Zichichi) (Editrice Compositori, Bologna, 1973), p. 433.

10) M.S. Plesset, Phys. Rev. 41,278 (1932).

11) J. Kogut and L. Susskind, Phys. Rev. Letters 34,767 (1975).

12) H.J. Schnitzer, Hyperfine splitting of ground state charmonium, Brandeis University Preprint (June 1975).

# ON THE FORMULATION OF RELATIVISTIC HADRON COUPLINGS*

A. N. Mitra
Dept. of Physics and Astronomy, University of Delhi, India

The problem of relativistic structure of hadronic matrix elements has acquired renewed interest in recent times, in the wake of the so-called Melosh transformation.[1] The Melosh approach has, among other things, focussed attention on the need for a broader logical and conceptual framework for a clearer understanding of several empirical recipes which have proved highly successful in accounting for many hadronic phenomena. Especially interesting is the natural understanding it provides for the quark recoil effect, or the $^3P_0$ $Q\bar{Q}$ pair mechanism, which provides a successful description of s-wave decays, anti-$SU(6)_W$ signs, etc.[2] On the other hand, the Melosh approach does not, without additional input (by way of free parametrization), provide a quantitative handle on hadronic matrix elements. In other words, while the Melosh theory provides formally indentical algebraic structures to what the more empirical approaches yield, its correlative powers none the less remain severely limited. Therefore, this theory does not by itself obviate the need for more concrete theories which, while respecting the Melosh features, should nevertheless make unambiguous and verifiable predictions on a wide variety of hadronic phenomena without extra parametrization along the way.

The object of this paper is to examine the possibilities for a more comprehensive basis for the formulation of hadron couplings. For this purpose, it is good to recognize that one must look well beyond the trad-

---

* This paper was submitted to the Workshop by Professor A. N. Mitra, who was in the end unable to be present at this Meeting.

itional radiation quantum (RQ) hypothesis within the additivity assumption,[3] and the SU(6)xO(3) framework for classification[4] which has proved broadly successful in making coupling predictions, but only with the (i) additional assumption of quark recoil effect,[5] (ii) explicit ansatz on the choice of the radiation quantum for purely mesonic transitions and (iii) a fair degree of parametrization of the form factors.[6,7] In overcoming the limitation of parametrization within the above framework, the relativistic harmonic oscillator (h.o.) model[8] has proved to be a big unifying step theoretically, while broadly conforming to the data. It even predicts a recoil effect, but the magnitude of the effect is not adequate for explaining the (L-1) wave couplings. Therefore the RQ hypothesis with h.o. wave functions - the FKR approach - does not quite suffice, either theoretically or experimentally, as a satisfactory base for generating hadron couplings.

A more promising starting point is the so-called quark-pair-creation (QPC) hypothesis which was also proposed several years ago,[9] though its formal advantages over the RQ hypothesis which would accrue from a more explicit use of the wave functions of the quark composites involved, had not been fully exploited in the initial proposals.[9] It is only in more recent times, presumably stimulated by the successful applications of h.o. wave functions in predicting hadronic decays and other phenomena, that the fuller possibilities of the QPC hypothesis have been brought to sharper focus.[10] In consonance with h.o. wave functions, the latter (i) incorporates the recoil effect, (ii) is indifferent to the choice of the RQ for purely mesonic transitions and (iii) successfully overcomes the limitations of parametrization of form factors. Thus the QPC hypothesis appears as an intrinsically more attractive candidate than RQ, for providing a more comprehensive framework for hadron couplings.

Another major (but independent) dimension to the problem of hadron couplings is represented by the nature and scope of their relativistic formulation. In this respect both 3-dimensional (non-covariant)[11] and 4-dimensional (covariant)[12] types of formulation have been proposed, but one must not conclude a priori that the 4-dimensional form is necessarily superior to the 3-dimensional one, especially after the developments connected with the Melosh approach. Theoretically, the success of a covariant formulation is linked with the probability interpretation of the Bethe-Salpeter wave functions of the quark composites, which is beset with "problems" despite the very notable efforts made recently by Kim and collaborators.[13] At the observational level too, the lack of success of some covariant models, warrants a closer look at non-covariant formulations, especially one due to Licht and Pagnamenta[11] which successfully explained the e.m. form factor of the proton.

We believe that the above considerations provide useful material for the formulation of a more comprehensive base for the description of hadron couplings with wide correlative powers, but it is first necessary to make a more detailed scrutiny of the RQ versus QPC hypotheses on the one hand, and the virtues of the 3-dimensional versus 4-dimensional relativistic formulations on the other. Unfortunately, an optimal formulation from this twin point of view cannot be based on a successful comparison with the data which are (at best) not very sensitive to the above different assumptions in the low mass region of resonances, and not adequately available for high mass (and high spin) ones. One therefore needs more powerful criteria going well beyond the mere correlative powers of a theory.

Our approach to the problem of choice of alternatives is based on the following considerations bearing on the behaviour of the resonant mass (M) as a function of the total excitation quantum number (N). The h.o. model gives the (experimentally well verified) result that the real part of $M^2$ varies linearly with N, but does not make a corresponding prediction on the variation of its imaginary part, i.e. the total width ($\Gamma_{tot}$) with N. Though the empirical data on the total widths of several resonances (especially $\Delta$ and N) strongly suggests a linear variation, there does not seem to exist any formal theory so far to limit the dependence of $\Gamma_{tot}$ on N. We now seek to link our quest for a satisfactory formulation of hadron couplings to the requirement that the rise of $\Gamma_{tot}$ with N be <u>less than exponential</u> and not necessarily linear (as suggested by the data). As we shall see in the following, even this relatively mild assumption appears adequate for a rather unique choice among the <u>four</u> alternatives listed above, viz (a) QPC versus RQ and (b) 4-dimensional versus 3-dimensional relativistic structures. The simple answer turns out to be that only the QPC with a 3-dimensional formulation passes the above test while the other three predict exponential rise of $\Gamma_{tot}$ with N. The ingredients used for the derivation of this result are (i) h.o. wave functions and (ii) a simple inclusiveness ansatz for the estimation of total widths based on the h.o. prediction of quantized masses at intervals of $M^2 \approx 1\mathrm{GeV}^2$ and the simulation of multibody decays by successive two-body ones.

We now describe the essential features of the calculations, the details of which are given elsewhere.[14] It is adequate to consider the quarks as "scalars" with respect to SU(6) degrees of freedom and to confine attention only to $M \rightarrow MM$ transitions which contain all the features of physical and mathematical interest. Under the RQ hypothesis the form factor is the overlap integral of <u>two</u> wave functions corresponding only to the "composite" hadrons, while the QPC assumption requires the overlap integral to include the wave functions of all the three hadrons symmetrically. In the N.R. limit, the QPC matrix element for the transition $M_a \rightarrow M_b + M_c$ has the form[10,13,14]

$$V_B = G_B\delta(-\vec{p}_a + \vec{p}_b + \vec{p}_c)\int d\vec{q}\phi_b^+(-\vec{q} + \sqrt{2}\vec{p}_a - \sqrt{2}\vec{p}_b)\phi_c^+(-\vec{q} - \sqrt{2}\vec{p}_b)\phi_a(\vec{q}) \quad (1)$$

where the external momenta are $\vec{p}_{a,b,c}$ and the internal ones are given by the arguments of the respective wave functions. The momentum normalizations are as in FKR.[8] This form is appropriate for a 3-dimensional, non-covariant, Licht-Pagnamenta (LP)[11] type relativistic formulation in the Breit frame between $M_a$ and $M_b$. The recipe which is explained elsewhere[15] consists in the following replacements (corresponding to the Breit frame momenta $\pm\vec{p}$) for the external momenta appearing in the arguments of the wave functions in (1):

$$\vec{p}_a \rightarrow \vec{p}_a' = \frac{\vec{p}_a m_b}{E_{bp}}, \qquad \vec{p}_b \rightarrow \vec{p}_b' = \vec{p}_b \frac{m_a}{E_{ap}} \quad (2)$$

$$E_{ap}^2 = p^2 + m_a^2, \text{ etc.}$$

There is also an overall Lorentz contraction factor $(m_a m_b/E_{ap}E_{bp})$ multiplying the matrix element so that the net relativistic matrix element for the Breit frame under the QPC assumption becomes:

$$\bar{V}_B = G_B\delta(-\vec{p}_a + \vec{p}_b + \vec{p}_c)\int d\vec{q}\phi_b^+(-\vec{q} + \sqrt{2}\vec{p}_a' - \sqrt{2}\vec{p}_b') \times \phi_c^+(-\vec{q} - \sqrt{2}\vec{p}_b')\phi_a(\vec{q})m_a m_b/E_{ap}E_{bp} \quad (3)$$

An alternative form of (1) which is amenable to a 3-dimensional relativistic formulation corresponding to the c.m. frame of the final mesons $M_b$ and $M_c$ is obtainable in terms of the momenta $\vec{p}_b$ and $\vec{p}_c$:

$$V_B^{c.m.} = G_B\delta(-\vec{p}_a + \vec{p}_b + \vec{p}_c)\int d\vec{q}\phi_b^+(-\vec{q} + \sqrt{2}\vec{p}_c)\phi_c^+(-\vec{q} - \sqrt{2}\vec{p}_b)\phi_a(\vec{q}) \quad (4)$$

An analogous L-P type argument in this case leads to the following modification on the external momenta in Eq.(4), appropriate to an alternative 3-dimensional relativistic structure:

$$\vec{p}_b \rightarrow \vec{p}_b'' = -\vec{k}\frac{m_c}{E_{ck}}, \qquad \vec{p}_c \rightarrow \vec{p}_c'' = +\vec{k}\frac{m_b}{E_{bk}} \quad (5)$$

together with an overall Lorentz contraction factor multiplying the N.R. matrix element (4). In this case the resultant form of the relativistic matrix element under the QPC assumption is:

$$\bar{V}_B^{c.m.} = G_B\delta(-\vec{p}_a + \vec{p}_b + \vec{p}_c)\frac{m_b m_c}{E_{bk}E_{ck}}\int d\vec{q}\phi_b^+(-\vec{q} + \sqrt{2}\vec{p}_c'')\phi_c^+(-\vec{q} - \sqrt{2}\vec{p}_b'')\phi_a(\vec{q}) \quad (6)$$

For a 4-dimensional co-variant structure of the matrix element, the most promising formulation of normalized 4-dimensional wave function which successfully overcomes the problem of suppression of time-line excitations, is one due to Lipes[12)] and K-N.[13)] This formulation is based on the use of h.o. wave functions, whose ground state form for a meson of 4-momentum $P_\mu$ is

$$\psi_o(q_\mu, \hat{p}) = \pi^{-1}\exp[-\tfrac{1}{2}q^2(\hat{p}) - \tfrac{1}{2}(q.\hat{p})^2] \quad (7)$$

$$q_\mu(\hat{p}) = q_\mu - q.\hat{p}\ \hat{p}_\mu, \quad (8)$$

while the excited states are described by multiplying Eq.(7) with suitable spherical harmonics and Laguerre functions of the effective 3-momentum $q_\mu(\hat{p})$. The covariant overlap integrals (denoted by $V_B$) in the QPC case can now be read directly from Eq.(1) (or Eq.(4)) through the mere interpretation of the various 3-momenta as 4-momenta and using the above forms of the wave functions. The non-covariant forms (3) or (6), on the other hand, are not necessarily limited by the choice of h.o. wave functions (since the problem of normalization of 3-dimensional wave functions is much less serious than that of 4-dimensional ones). However, in practice mainly h.o. wave functions are useful in the non-covariant case, not only because of their experimental success, but also for providing a parameter-free handle for calculations.

It is easy to deduce the corresponding forms of the matrix elements under the PQ hypothesis from the above expressions for QPC matrix elements by merely suppressing the internal wave function of the specified RQ, and exhibiting it as an external wave function with the appropriate momentum.[14] Thus with one of the outgoing mesons ($M_c$) as the RQ, the corresponding matrix element in the N.R. case is deducible from (1) as

$$V_A = G_A\delta(-\vec{p}_a + \vec{p}_b + \vec{p}_c)\phi_c^+(p_c)\int d\vec{q}\phi_b^+(-\vec{q} + \sqrt{2}\vec{p}_a - \sqrt{2}\vec{p}_b)\phi_a(q) \tag{9}$$

A similar expression holds when the initial meson ($M_a$) is regarded as the RQ. Relativistic extensions from Eq.(9) can be carried out in a non-covariant manner as in Eq.(3) for the Breit frame or in a 4-dimensional covariant manner according to the considerations following Eq.(7). (We denote these expressions by $\bar{V}_A$ and $\tilde{V}_A$, respectively).

For a comparison among the four alternatives $\bar{V}_A$, $\tilde{V}_A$ and $\bar{V}_B$, $\tilde{V}_B$ representing the RQ versus QPC hypothesis and non-covariant versus covariant relativistic formulations, we have chosen to study the <u>total decay widths</u> ($\Gamma_{tot}$) for the following two types of transitions:

(I) Radially (n) excited meson $M_a$ decaying into a pair $M_b, M_c$ with masses $M_1^2 \approx n_1$, $M_2^2 \approx n_2$ where $n_1$ and $n_2$ are integers, consistent with the h.o. constraint ($n \leq n_1 + n_2$).

(II) Orbitally (L) excited meson $M_a$ decaying into a pair $M_b, M_c$ of lower orbital excitations ($\ell_1$ and $\ell_2$) consistent with the h.o. constraints on their masses ($\frac{1}{2}L \geq \ell_1 + \ell_2$).

To evaluate $\Gamma_{tot}$ in terms of the contributions of individual channels ($n \to n_1 n_2$) our basic assumption is to span all the partition integers $n_1$ and $n_2$ which incorporate the law of masses provided by the h.o. model, i.e. $\Delta M^2 \approx 1\text{GeV}^2$. Implicit in this mechanism is the assumption that multibody ($\geq 3$) decays are simulated by all possible two-body modes characterized by the integers $n_1$ and $n_2$ corresponding to successively higher masses. Thus for the radially excited (n) decays, the total width n is expressible in terms of the partial widths $\Gamma(n \rightarrow n_1 n_2)$ as

$$\Gamma_n = \sum_{n_1+n_2 \leq n} \left(\frac{n!}{n_1!n_2!(n-n_1-n_2)!}\right)\Gamma(n \rightarrow n_1 n_2) \tag{10}$$

where the restriction $n_1 + n_2 \leq n$ follows from the asymptotic mass relations

$$M_1^{\,2} \approx n_1, \; M_2^{\,2} \approx n_2, \; M^2 \approx 2n \tag{11}$$

as well as the reality condition on the decay momentum. Since the asymptotic partial width depends effectively on the single variable $\nu = n_1 + n_2$, Eq.(10) simplifies to

$$\Gamma_n = \sum_{\nu \leq n} \frac{2^{\nu} n!}{\nu!(n-\nu)!} \Gamma(n \rightarrow \nu) \tag{12}$$

This formula which represents our <u>inclusiveness ansatz</u> for the estimation of $\Gamma_n$ holds under the assumption of quantized $(\text{mass})^2$ units given by the h.o. model. A formula similar to (10) holds also for the casse II representing the decay of an L-excited meson, with the replacements:

$$n_1 \rightarrow \ell_1, \; n_2 \rightarrow \ell_2, \; n \rightarrow \tfrac{1}{2}L \tag{13}$$

However, since the partial width $\Gamma(L \rightarrow \ell_1 \ell_2)$ is now proportional to $(\ell_1!)^{-1}$ x $(\ell_2!)^{-1}$ times a function $Z_{L\ell}$ of $1 = 1_1 + 1_2$, the formula analogous to (12) is now:

$$\Gamma_L = \sum_{\ell \leq \frac{1}{2}L} \frac{(2\ell)!}{(\ell!)^4} \frac{(\frac{1}{2}L)!}{(\frac{1}{2}L-\ell)!} Z_{L\ell} \tag{14}$$

These formulae provide the necessary basis for the comparison of the four candidates ($\bar{V}_A$, $\bar{V}_B$, $\tilde{V}_A$, $\tilde{V}_B$) through their dynamical effects on the reduced partial widths $\Gamma(n \rightarrow \nu)$ and $Z_{L\ell}$. The calculations are lengthy but straightforward, and the results for radially excited decays are summarized in the following table for $(72\pi^2 G^{-2} \Gamma_{tot})$ in the asymptotic region[14]:

| | non-covariant | covariant |
|---|---|---|
| QPC | $\sqrt{\frac{n}{e}} \times e^{-0.61n}$ | $\frac{0.1}{n^3} e^{0.17n}$ |
| RQ | $9\pi \times 3^n$ | $\sqrt{\frac{2\pi}{n}} e^{0.37n}$ |

These figures indicate that the RQ hypothesis is definitely disfavoured by the criterion of less-than-exponential[11)] rise of $\Gamma_{tot}$, while the QPC hypothesis passes the test. The same criterion suggests that the relativistic version is more promising at the 3-dimensional non-covariant level than the more elegant, covariant form, even though in the latter case the small numerical coefficient of $n$ in the exponential may give the illusion of a mild initial rise with $n$. Similar results are indicated for the decays of orbitally (L) excited states as well. Thus for the two principal forms of excitations (radial and orbital) the feature of "less-than-exponential" rise of $\Gamma_{tot}$ seems to hold only under the QPC assumption with a three-dimensional relativistic form of the matrix element.

To get a numerical idea of the significance of the exponential rise of $\Gamma_{tot}$ with the excitation quantum number, some estimates have recently been made of $\Gamma_{tot}$ for successive radial excitations of the $\rho$- meson through various available channels such as PP, VP, VV, $\bar{B}$ B, $\bar{B}$ B*, $\bar{B}$* B*, together with their respective excited states, using the different alternatives listed above for the coupling scheme.[15)] All the 3 cases which predict exponential rise of $\Gamma_{tot}$ with $n$ were found to yield magnitudes for $\Gamma_{tot}$ exceeding several GeV even in the modest mass region of 3-4GeV.[15)] These would clearly appear to be quite unrealistic in terms of the estimates available from the PDG tables.[16)] On the other hand, the QPC version with non-covariant relativistic structure ($V_B$) turned out to yield rather modest figures eventually tapering off to zero for high values of $n$. Since any estimate of $\Gamma_{tot}$ howsoever realistic, is likely to be an underestimate, our model for $\Gamma_{tot}$ cannot also be free from this limitation. However, it is only the $\bar{V}_B$ version which seems to leave scope for further increase of $\Gamma_{tot}$ without the danger of an exponential growth of this quantity.

Thus our simple criterion seems to narrow down the choice of alternatives for the formulation of relativistic hadron couplings to a significant extent. A priori, all the four possibilities considered in the paper have comparable predictive power which stems primarily from the use of h.o. wave functions and the $SU(6)_W \times O(3)$ model for couplings. However, all of them do not seem to conform to additional theoretical requirements with observable consequences. The QPC version enjoys an intrinsically higher degree of universality (and hence correlative power) than RQ, since the former provides the recoil term without extra parametrization, and does not also need any extra assumption for the choice of the radiation quantum for mesonic transitions. That the QPC hypothesis is compatible with our chosen constraints on $\Gamma_{tot}$ lends an added virtue to a theory which has a greater theoretical appeal than its RQ counterpart.

As for relativistic effects it is perhaps somewhat of an aesthetic disappointment that the covariant formulation is disfavoured by our criterion on $\Gamma_{tot}$ compared with a non-covariant version,[17)] but the comparison with data should help dispel any impression of an automatic superiority of the covariant model over a non-covariant one.

Regarding the performance of QPC relative to RQ, it appears from the work of Le Yaouanc et al.[17)] that the former is no less satisfactory than the latter, at least in the resonance region up to 2GeV. In addition, the magnitude of the recoil effect predicted by this model is quite adequate for an understanding of (L-1) wave decay features unlike the RQ model of FKR.[8)] The real discrimination should be expected from higher resonance features, as indicated by some recent calculations on the widths of higher radial excitations.[15)] Several related investigations to further test the QPC hypothesis with non-covariant relativistic structure are currently in progress.

I am grateful to Indrakumari for permission to quote from her calculations prior to publication.

## REFERENCES

1. H. J. Melosh, Phys. Rev. D9, 1095 (1974).

2. See, e.g. J. L. Rosner in Proc. XVII Intl. Conf. on High Energy Physics, London, July 1974.

3. C. Becchi and G. Morpurgo, Phys. Rev. 149, 1284 (1966).

4. R. H. Dalitz in Proc. XIV Intl. Conf. on High Energy Physics, Berkeley, 1966.

5. A. N. Mitra and M. Ross, Phys. Rev. 158, 1630 (1967).

6. For a review of the earlier background on this subject, see, e.g. J. J. Kokkeedee, the Quark Model (W. A. Benjamin and Sons, New York, 1969).

7. For a more recent review, see, e.g. A. N. Mitra, Supermultiplet Interactions among Hadrons and Resonances, Delhi University Report 1973 (unpublished).

8. R. P. Feynman, M. Kislinger and F. Ravndal, Phys. Rev. D3, 2706 (1971) referred to as FKR.

9. L. Micu, Nucl. Phys. B10, 521 (1969).
R. Carlitz and M. Kislinger, Phys. Rev. D2, 336 (1970).
E. Colglazier and J. Rosner, Nucl. Phys. B27, 349 (1971).

10. A. Le Yaouanc, L. Oliver, O. Pene and J. C. Raynal, Phys. Rev. D8, 2223 (1973).

11. A. Licht and A. Pagnamenta, Phys. Rev. D3, 1150 (1970).

12. K. Fujimura, T. Kobayasi and M. Mamiki, Prog. Theo. Phys. 43, 73 (1970)
R. Lipes, Phys. Rev. D5, 2849 (1970).

13. Y. S. Kim and M. Noz, Phys. Rev. D8, 3521 (1973), and subsequent papers.

14. Indrakumari and A. N. Mitra, Delhi Univ. Preprint (1975) - to be published.

15. Indrakumari and A. N. Mitra, 1975 (to be published).

16. Particle Data Group, in Phys. Letters, April 1974.

17. See, e.g. Le. Yaouanc et al., Phys. Rev. D11 , 5, 1272 (1975).

# QUARK DYNAMICS IN THE SLAC BAG MODEL*

R. C. Giles

Stanford Linear Accelerator Center
Stanford University, Stanford, California 94305

## I. INTRODUCTION

As has been evident in most of the discussions of this workshop, there is some advantage to the phenomenological description of hadronic structure and interactions in terms of quark constituents. However, we have yet to produce a simple fundamental theory of quark interactions from which this phenomenology arises. Perhaps the aspect of quark dynamics that most challenges our imagination (and our mathematical capacities) is the apparent absence of free quark states.

In its barest form the puzzle is that, although the successes of the nonrelativistic quark model spectrum, the quark parton model of deep inelastic processes, Zweig's rule, and most of the rest of the usual quark phenomenology suggest a picture of quarks as weakly interacting constituents of hadrons, the absence of asymptotic quark states requires strong binding. Actually, the restriction is more severe: there exist no exotic (triality nonzero) hadrons. This requires not only strong binding but what may be properly called "quark confinement": that is the confinement of both quarks <u>and</u> their quantum numbers.

It is generally appreciated that the apparent contradiction between these aspects of quark dynamics may be resolved in a model which has the property that quarks interact weakly at short distances and strongly only at large distances. A realization of such dynamics is very difficult in perturbative quantum field theory. Only in the case of asymptotically free gauge theories does perturbation theory give a reliable guide, and then only to the short distance behavior of the interactions. The strong coupling problem which arises at distances on the order of a hadron radius cannot be solved perturbatively.

---

*Work supported by the U. S. Energy Research and Development Administration.

In this talk, we concentrate on another, nonperturbative, variety of field theoretic models—the "bag" models—with particular attention to the "SLAC Bag"[1] or "bubble" model. Before plunging into a detailed discussion of this model, it is perhaps useful to consider some more general features of "bags". In this talk, I will characterize three theories as "bag" models. These are the MIT bag,[2] the bubble, and the Nambu string.[3] The basic advantage of the description of hadrons as bags is that both the long range forces which account for strong binding and the hadron of quark motion at short distances is incorporated at the outset. The most troublesome aspect of these models is that no fully quantized theory of bags has been constructed. The simple picture of quark dynamics and numerical results for the hadron spectrum depend on a semi-classical analysis. We will see that the quantum effects which are neglected in this approach are essential for the understanding of form factors and of hadronic interactions. Whether a satisfactory quantum theory of bags can be constructed remains a crucial open question.

The semi-classical picture of quark bound states in bag models is one in which quarks are bound to an extended, coherent excitation of the vacuum. Inside the space-time region of this excitation (the "bag") quarks move nearly freely. Outside this region, quarks have a large or infinite effective mass and a correspondingly small penetration. The various "bag" models are distinguished, in four space-time dimensions, by the spacial dimensionality of the vacuum excitation upon which quarks are constrained to move: string[4] (one-dimension), bubble (two-dimensions), and bag (three-dimensions). The geometric configuration of the bag is a dynamic variable (though not, in the case of the MIT bag, and independent one). A constant energy density is associated with the volume of the bag. The bag configuration is determined by the local balance of field pressure against this constant tension. In this view, as suggested by Prof. Joos, the bag models afford perhaps the simplest generalization of a potential model in which the "potential" exchanges momentum as well as energy with quarks and is itself possessed of dynamical structure. Semi-classically, bag models are formulated in a manifestly Lorentz covariant way.

The bag models take into account strong quark binding in a simple way. However, quark confinement (absence of triality nonzero states) is achieved only after the introduction of $SU(3)_{color}$ degrees of freedom and (weakly) coupled color gauge fields within the bag. Quark confinement appears less natural here than the confinement which is conjectured to arise in asymptotically free gauge theories due to "infrared slavery" (at present more accurately referred to as "infrared confusion"). This bit of inelegance is, perhaps, the price we pay for being able to calculate anything (albeit semi-classically).

Finally, as we shall see in great detail in the case of the bubble, bag models can be formulated semi-classically in two ways. First, there exists a "canonical" formulation in which the dynamic degrees of freedom are those describing the geometry of the bag and the quark fields defined on it. The dynamics of the system is determined from an action principle (along with possible subsidiary conditions on the fields). It has also been shown[5,6,7)], however, that bag-like states can form in field theories. It is found that certain conventional classical interacting field theories possess solitary wave solutions which, in an appropriate strong coupling limit, are described by the same system of equations as are the corresponding bags[8)]. In these field theories, quarks are not absolutely bound to the bag—there exists a finite energy threshold for the production of free quarks. With a choice of sufficiently large couplings, this threshold may be set arbitrarily high without affecting low-energy bag dynamics. Only in the "limit" of infinitely strong coupling does the solitary wave go over exactly to the bag.

This suggests an alternative approach to the hitherto futile (in four-dimensions) attempt to quantize bags. The quantum field theories corresponding to those classical theories which have bag-like solutions are well defined. It may be possible to find states of the quantum field theory in the strong coupling limit which correspond to the classical solitary wave. Indeed, it was an approximate variational calculation of the energy of a coherent state in quantum field theory which led Bardeen, Chanowitz, Drell, Weinstein, and Yan (BCDWY) to the bubble theory. This line of investigation is actively being pursued at SLAC by Drell[9)], Weinstein, and Yankielowicz[10)].

In this talk, we will discuss some of the above ideas in the context of the bubble model of BCDWY. Beginning with a conventional classical field theory of interacting quarks and a scalar, we show that in the case of strong coupling extended thin shell bound states form. We extract from the field theory a classical "canonical" theory of these bubbles. We discuss some exact and approximate semi-classical solutions to the bubble equations of motion and their implications for the hadronic spectrum in the BCDWY model. Finally we discuss some of the problems of quantization of the canonical theory and its qualitative modifications to the semi-classical picture of the states. In particular, we [11)] discuss the quantum theory of a single bubble in 2+1 space-time dimensions where the classical theory is exactly solvable and its quantization is therefore straightforward.

## II. THEORY OF BUBBLES

We begin by considering the binding of a single quark species. As we shall see, the generalization to several species (colors) is straightforward. The BCDWY model for the binding of a single quark species is developed from the field theory defined by the Lagrangian:

$$L = \frac{1}{2}(\partial\sigma)^2 - \lambda(\sigma^2 - f^2)^2 + \bar{\psi}(i\partial\!\!\!/ - G\sigma)\psi$$

whose Hamiltonian is

$$H = \int dx \, \frac{1}{2}\left(\frac{\partial\sigma}{\partial t}\right)^2 + \frac{1}{2}(\nabla\sigma)^2 + \lambda(\sigma^2 - f^2)^2 + \psi^+(-i\alpha\cdot\nabla + G\sigma\gamma^o)\psi$$

This is a theory of a Dirac field $\psi$ interacting, via the Yukawa coupling, with a quartically self-coupled, neutral scalar field, $\sigma$. The theory is characterized by three coupling constants: G, $\lambda$, and f.

This Lagrangian is symmetric under the discrete transformation:

$$\sigma \to -\sigma$$
$$\psi \to \gamma_5\psi$$

The classical potential of the $\sigma$ field has symmetric minima at $\sigma = \pm f$. One therefore expects that, in the corresponding quantum theory, $\gamma_5$ reflection symmetry will be spontaneously broken and that the $\sigma$ field will assume a vacuum expectation value $|\langle\sigma\rangle| = f$, which we choose, by convention, to be +f. In a perturbative approach one would then conclude that this theory is one of interacting quarks of mass $M_Q = Gf$ and scalar mesons of mass $M_\sigma = \sqrt{8\lambda}f$. We consider a limit of coupling constants in which both of these masses are "large".

It is easy to construct a semi-classical argument that the lowest lying quark states need not have mass Gf. It is only in zeroth order perturbation theory that the scalar field is not free to decrease its value from f in the neighborhood of the quark, so as to reduce the total energy. The particle-like excitations of the semi-classical BCDWY theory are formed in just this way.

The "semi-classical" field equations which we use to discuss the BCDWY theory consist of the (one-particle) Dirac equation for $\psi$ in the presence of a classical $\sigma$ field:

$$(i\partial\!\!\!/ - G\sigma)\psi = 0 \tag{1}$$

where the wave function must be normalized to unit charge,

$$Q = \int dx\, \psi^{+}\psi = 1$$

and of the classical equation for the $\sigma$ field in the presence of a fermion source:

$$-\partial^2\sigma + 4\lambda\sigma(f^2-\sigma^2) = G\bar{\psi}\psi \qquad (2)$$

In the "static" case, $\sigma = \sigma(\vec{x})$, $\psi = \psi(\vec{x})\, e^{-iEt}$ these reduce to:

$$(-i\alpha\cdot\nabla + G\sigma\gamma^{o})\,\psi(\vec{x}) = E\psi(\vec{x})$$
$$\nabla^2\sigma + 4\lambda\sigma(f^2-\sigma^2) = G\bar{\psi}\psi$$

The differential equations (1) and (2) are the classical Euler-Lagrange equations of the theory. The system is "semi-classical" in the sense that $\psi$ is interpreted as if it were a single-particle Dirac wave function: it is normalized to unit charge and negative energy fermion states are to be given the Dirac interpretation as positive energy antifermions. We note that the Dirac equation is one with a scalar potential, so that no Klein paradox arises—the distinction between positive and negative energy states is always unambiguous. Thus, the prescription by which we define a "semi-classical" theory is also unambiguous. The normalization of the fermion charge to 1 and the interpretation of the negative energy sea as antiparticles arise naturally in the work of BCDWY where these semi-classical equations are derived from the quantum field theory via an approximate variational technique.

The mechanism by which low mass quark bound states can form is most clearly evident in the solution to the static field equations in one dimension. Taking the representation of the gamma matrices:

$$\gamma^{o} = \begin{pmatrix} 1 & 0 \\ 0 & -1 \end{pmatrix} \qquad \alpha = \begin{pmatrix} 0 & 1 \\ 1 & 0 \end{pmatrix}$$

this solution is

$$\sigma(x) = f \tanh \sqrt{2\lambda}\, f(x-x_0)$$

$$\psi(x) = \frac{N}{\sqrt{2}} \left[\cosh \sqrt{2\lambda}\, f(x-x_0)\right]^{-G/\sqrt{2\lambda}} \begin{pmatrix} 1 \\ i \end{pmatrix}$$

where N is a normalization constant that insures Q=1, and $x_0$ is a constant.

One finds

$$E = 0$$

$$E_{total} = \frac{4}{3}\sqrt{2\lambda}\, f^3$$

$$\bar{\psi}\psi = 0$$

In a strong coupling limit defined by

$$G, \lambda \rightarrow \infty$$
$$f \rightarrow 0$$
$$G >> \lambda^{1/6}$$

and

$$\lambda^{1/6} f = \text{fixed} \quad ,$$

this is clearly a one quark state of much lower energy than the usual free quark.

There are several aspects of this one-dimensional solution which point toward more general features of the theory. First, because $\bar{\psi}\psi$ vanishes, the $\sigma$ field equation is actually independent of $\psi$. The above solution for $\sigma$ is the well-known "kink" solution of the spontaneously broken quartic scalar theory in one dimension. The dynamics of the scalar field is determined primarily by its self-coupling, rather than by its coupling to fermion sources. This will remain true in higher dimensions. The width, D, of the transition region of the $\sigma$ field is on the order of the $\sigma$ Compton wavelength, which will always be small compared to $(1\ \text{GeV})^{-1}$.

Perhaps the most striking feature of the solution is that the Dirac energy is small even though the Dirac wave function is very sharply peaked. Intuition based on the quantum mechanics of bosons would suggest that the energy should be comparable to the dominant Fourier components of the wave function—on the order of the bare quark mass. This intuition need not be correct because the Hamiltonian is linear, rather than quadratic, in the quark momentum operator. This point is fully discussed in Ref. 1.

In higher dimensions, as discovered by BCDWY, the low-lying bound states analogous to the one-dimensional kink have the form of finite domains within which $\sigma$=-f and outside of which $\sigma$=+f. The transition of the $\sigma$ field between these values is very sharp and takes place in a thin shell about some closed surface in space (Fig. 1). Quarks can be trapped on this domain boundary as they are on the kink. We refer to such states as "bubbles". In general, a bubble's surface may vary in time. Thus, the most natural description of a bubble is as a domain in space-time whose boundary surface is time-like hypertube (Fig. 2).

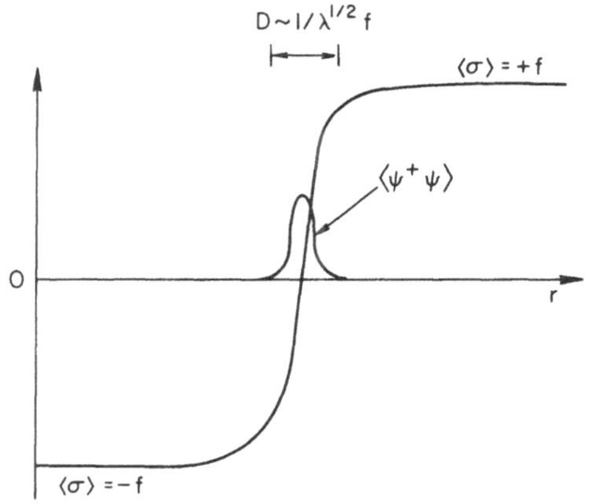


Fig. 1 A schematic representation of a bubble state.

We discuss a general approximation scheme which affords a characterization of all bubble solutions to the Euler-Lagrange equations. Our approximate solutions become exact in the infinitely strong coupling limit. The approximations we use in this discussion will give physical quantities to lowest order in a small parameter which we may denote schematically as "D/R". Here, D is a length on the order of the Compton wavelength of the quark or the meson and R is on the order of the smallest radius of curvature of the bubble surface. Thus, D/R is the ratio of the thickness of the shell to its size and, as we shall see, vanishes in the strong coupling limit.

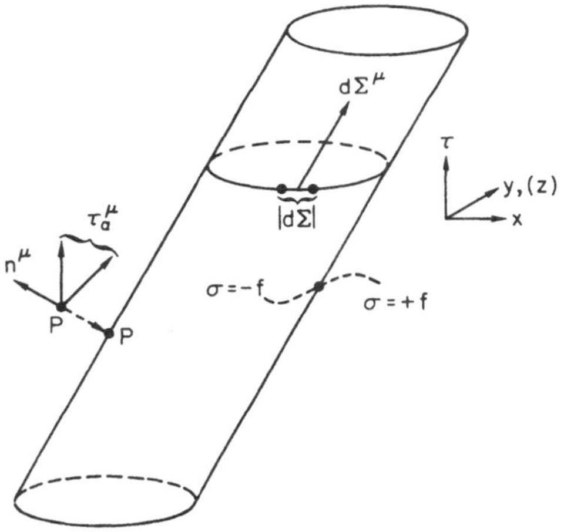


Fig. 2 A hypertube.

Our procedure is as follows:

(1) We assume that the desired solution to the field equations is a bubble of some undetermined shape. The $\sigma$ field equation may be solved approximately for any such configuration.

(2) We solve the Dirac equation approximately in the presence of this $\sigma$ field.

(3) Finally, we derive a self-consistency condition which guarantees that the next order corrections to this approximate solution are, in fact, small.

We begin with the assumption that the field configuration will be that of a bubble (Fig. 2). In n-dimensional Minkowski space, the bubble surface, $\sigma(x)=0$, is an n-1 dimensional hypersurface which we can parametrize by n-1 "internal coordinates", $u^\alpha$.

$$x^\mu = R^\mu(u^\alpha) \qquad \begin{array}{l} \alpha=0,\ldots,n-2 \\ \mu=0,\ldots,n-1 \end{array}$$

Because the fields are expected to have nontrivial space-time dependence only in a thin shell about this surface, it is convenient to adopt (non-cartesian) coordinates $(u^\alpha, \xi)$ centered about it.

$$x^\mu(u^\alpha,\xi) = R^\mu(u^\alpha) + \xi n^\mu(u^\alpha)$$

where $n^\mu(u^\alpha)$ = outward unit normal at $R^\mu(u^\alpha)$.

The coordinates $(u^\alpha, \xi)$ are well defined only within a distance on the order of one radius of curvature away from the surface. We assume that the radii of curvature of the bubble surface are always large compared to D. This assumption has no effect whatsoever on the spectrum of low-lying excitations of the theory in the strong coupling limit. By increasing G and $\lambda$, D may be made arbitrarily small without affecting either the spectrum or the surface geometry[12)].

In the new coordinate system, we can write the gradient:

$$\partial_\mu = \partial_{\parallel\mu} - n_\mu \frac{\partial}{\partial\xi}$$

where $\partial_{\parallel\mu}$ is the "tangential" gradient which, though it depends on $\xi$, involves only differentiations with respect to the $u^\alpha$, and is tangent as a vector to the surface.

Consider the field equation for $\sigma$. Our first approximation to $\sigma$ must be a function that satisfies the "largest" part of Eq. (2) near the surface. Because $\sigma$ makes its transition from -f to +f in a distance of order D, we expect

$$\frac{\partial\sigma}{\partial\xi} \sim \frac{1}{D} f$$

while

$$\partial_{\|}\sigma \sim \frac{1}{R} f \ll \frac{1}{D} f$$

We also anticipate that, in analogy to the one-dimensional case, the fermion source term will be relatively unimportant in (2)—an assertion which must be verified later to insure self-consistency. Thus our first approximation to (2) in the neighborhood of the surface is:

$$\frac{\partial^2\sigma}{\partial\xi^2} + 4\lambda\sigma\,(f^2-\sigma^2) = 0 \tag{3}$$

This is the same as the equation for the kink of the one-dimensional theory. The solution of this equation which satisfies the boundary conditions and vanishes at $\xi=0$ is unique:

$$\sigma(x) = \sigma(\xi) = f \tanh \sqrt{2\lambda}\, f\xi \tag{4}$$

Next, we attempt to solve the Dirac equation (1) in the presence of this $\sigma$ field

$$\left[i\partial\!\!\!/_{\|} - i n\!\!\!/ \frac{\partial}{\partial\xi} - G\sigma(\xi)\right]\Psi = 0 \tag{5}$$

We construct an approximate solution valid as $G \to \infty$, using a technique similar to one invented by A. Chodos[2)] to derive boundary conditions for the Dirac field in the MIT bag theory. We expect that the Dirac wave functions will fall off exponentially away from the surface as $\sim e^{-Gf|\xi|}$. It is clear that such a $\Psi$ is not an analytic function of $1/G$ as $1/G \to 0$. However, we can attempt to factor out the essential singularity in $1/G$ and then expand its coefficient in $1/G$.

We write:

$$\Psi(u^\alpha,\xi) = N e^{GF(\lambda,\xi)}\left[\psi(u^\alpha,\xi) + \frac{1}{G}\psi_1(u^\alpha,\xi)\right] \tag{6}$$

where F and $\psi$ are independent of G, $\psi$ and $\psi_1$ are finite near $\xi=0$ as $G\to\infty$, $\psi+\frac{1}{G}\psi_1$ is the beginning of an expansion of the field in powers of $1/G$. Only the properties of the first term will be important, so we simply use the $\psi_1$ term to represent all higher corrections in $1/G$.

Substituting this form in the Dirac equation (5), we have

$$0 = -G\left[\sigma(\xi) + i\not{n}\frac{dF}{d\xi}\right]\psi$$

$$+\left(i\not{\partial}_{\|} - i\not{n}\frac{\partial}{\partial\xi}\right)\psi - \left[\sigma(\xi) + i\not{n}\frac{dF}{d\xi}\right]\psi_1 + 0\left(\frac{1}{G}\right) \qquad (7)$$

This equation must be satisfied order by order in 1/G. The equation for the coefficient of G is:

$$\left[f \tanh \sqrt{2\lambda}\, f\xi + i\not{n}\frac{dF}{d\xi}\right]\psi = 0$$

To have a nontrivial solution of this matrix equation for $\psi$ requires

$$\frac{dF}{d\xi} = \pm f \tanh \sqrt{2\lambda}\, f\xi$$

The requirement that F decrease with $|\xi|$ necessitates that we take the "-" sign above. We have

$$i\not{n}\psi = \psi$$

$$e^{GF} = \left[\cosh \sqrt{2\lambda}\, f\xi\right]^{-\frac{G}{\sqrt{2\lambda}}} \qquad (8)$$

The equation between the terms of order unity in (7) then becomes

$$\left(i\not{\partial}_{\|} - i\not{n}\frac{\partial}{\partial\xi}\right)\psi - \sigma(\xi)(1-i\not{n})\psi_1 = 0 \qquad (9)$$

Multiplying both sides by $(1+i\not{n})$ we find

$$\frac{\partial\psi}{\partial\xi} = -k\psi \qquad (10)$$

where

$$k \equiv \frac{1}{2}\,\partial_{\|\mu}(n^{\mu})$$

The quantity k depends on the surface geometry alone. In the next section we show that k is proportional to the local mean curvature of the surface.

At $\xi=0$, where the $\psi_1$ term in (9) vanishes we have:

$$(i\partial_{\|} + ki\not{n})\,\psi(u^{\alpha}, 0) = 0 \qquad (11)$$

This reduced Dirac equation involves only the behavior of $\psi$ on the surface. This equation and the equation of constraint, $i\not{n}\psi=\psi$, completely characterize the quark degrees of freedom of a bubble in the strong coupling limit.

Finally, we must show that the expressions we have obtained do constitute an approximate solution to the field equations. We will be led to a further equation of motion relating the surface geometry of the bubble and the distribution of quark energy-momentum on it. This condition is the generalization of the energy minimization principle used by BCDWY to determine the radius of the spherically symmetric status state in Ref. 1. The detailed derivation of this condition is straightforward but technically complicated[6]. We simply sketch the idea here.

Suppose we have fields $\sigma$ and $\Psi$ in the neighborhood of a bubble surface, such that

$$\begin{aligned} \sigma(\xi) &= f \tanh \sqrt{2\lambda}\, f\xi \\ \Psi &= N\left[\cosh \sqrt{2\lambda}\, f\xi\right]^{-G/\sqrt{2\lambda}} \psi(u^{\alpha}, \xi) \end{aligned} \tag{12}$$

where

$$i\not{n}\psi = \psi$$

$$\frac{\partial\psi}{\partial\xi} = -k\psi$$

$$\frac{1}{N^2} \equiv \int_{-\infty}^{\infty} d\xi \left[\cosh \sqrt{2\lambda}\, f\xi\right]^{-2G/\sqrt{2\lambda}}$$

These fields will be approximate solutions to the equations of motion in the strong coupling limit only if further corrections to them are of order D/R. These corrections may be estimated as follows: The action functional is expanded quadratically about the classical fields $\sigma$ and $\Psi$. In principle, the resulting quadratic function can be minimized, and shifts in the fields $\delta\sigma$ and $\delta\Psi$ and the corresponding change in the action $\delta S$ can be computed.

Because of the sharp gradients in the fields near the bubble surface, variations of the fields relative to this surface corresponds to very high frequency excitations which do not enter $\delta S$ to lowest order in D/R. It can be shown[6] that the only variations of the fields which can cause a finite shift, $\delta S$, are those which correspond to a motion of the surface and its associated fields as a whole. Only if the action is already stationary to order D/R under such variations will the fields $\sigma, \Psi$ be an approximate solution to the Euler-Lagrange equations.

That is

$$\frac{\delta}{\delta R^{\mu}(u^{\alpha})} \left[\int d^4x\, \bar{\Psi}(i\not{\partial}-G\sigma)\Psi + \frac{1}{2}(\partial\sigma)^2 - \lambda(\sigma^2-f^2)^2\right] = 0 \tag{13}$$

The lagrangian density in (13) is very sharply peaked in the neighborhood of the bubble surface. Thus the integral over $\xi$ can be performed to lowest order in D/R.

We have

$$S\sigma,\Psi = \int dx\, N^2 \left[\cosh\sqrt{2\lambda}\, f\xi\right]^{-2G/\sqrt{2\lambda}} \bar{\psi}\left[i\partial\!\!\!/_{\|} - i n\!\!\!/ \frac{\partial}{\partial\xi} - Gf\tanh\sqrt{2\lambda}\, f\xi\,(1-in\!\!\!/)\right]\psi$$

$$+\frac{1}{2}\left(-n_\mu\sqrt{2\lambda}\, f^2\operatorname{sech}^2\sqrt{2\lambda}\, f\xi\right)^2 - \lambda\left[f^2\operatorname{sech}^2\sqrt{2\lambda}\, f\xi\right]^2$$

$$\approx \int da\, d\xi\, N^2\left[\cosh\sqrt{2\lambda}\, f\xi\right]^{-2G/\sqrt{2\lambda}} \bar{\psi}(i\partial\!\!\!/_{\|} + kin\!\!\!/)\psi - 2\lambda f^4\operatorname{sech}^4(\sqrt{2\lambda}\, f\xi)$$

$$\approx \int_{\text{hypertube}} da\left[\bar{\psi}(i\partial\!\!\!/_{\|} + kin\!\!\!/)\psi - C\right]$$

where da = element of surface "area" on the hypertube and

$$C \equiv \int_{-\infty}^{\infty} d\xi\; 2\lambda f^4 \operatorname{sech}^4\sqrt{2\lambda}\, f\xi = \frac{4}{3}\sqrt{2\lambda}\, f^3$$

Thus we are led to a further equation of motion in the form of a "surface action principle":

$$0 = \delta\int da\left[\bar{\psi}(i\partial\!\!\!/_{\|} + kin\!\!\!/)\psi - C\right] \tag{14}$$

where the variation is to be performed over all possible bubble surfaces, $R^\mu(u^\alpha)$.

We note that the requirement that S be stationary under variations of the surface Dirac field $\psi$ leads to the correct surface Dirac equation (11). Thus, the dynamics of bubble states in the strong coupling limit can be completely described in terms of the geometric variables $R^\mu(u^\alpha)$, the surface Dirac field $\psi$, and the finite coupling C. The Dirac field obeys the constraint $in\!\!\!/\psi=\psi$. The equations of motion for $\psi$ and $R^\mu$ may be derived from the surface action principle (14).

These results may be easily understood physically. In the strong coupling limit, only a very special class of solutions to the field equations retain low energy. The requirement that their energy remain small forces these solutions to mimic, locally, the one-dimensional kink. The only remaining degrees of freedom are those which describe how such kinks are patched together continuously in space-time ($R^\mu(u^\alpha)$) and the quark distribution among them ($\psi(u^\alpha)$).

## III. BUBBLE DYNAMICS

The three equations (8), (11), and (14), completely characterize bubble solutions to the BCDWY field theory. In this section, we discuss some general properties of solutions to this system. The most natural language for the description of bubbles is that of the Riemannian geometry of surfaces. We will introduce some basic geometric notations and concepts in the following short discussion.

The surfaces whose geometry is of interest here are spacially closed n-1 dimensional timelike hypertubes imbedded in n-dimensional Minkowski space. Such a surface may be parametrized by n-1 "internal" coordinates, $\{u^\alpha\}$.

Surface:

$$x^\mu = R^\mu(u^\alpha)$$

Our notation will be such that $\alpha$, $\beta$, $\gamma$, $\delta$ run from $0, \ldots, n-2$, while $\mu$, $\nu$, $\lambda$, $\sigma$ run from $0, \ldots, n-1$.

The fundamental tensors characteristic of the surface geometry are defined in Table I.

Table I

Geometric Objects

| | |
|---|---|
| Tangent vectors: | $\tau^\mu_\alpha = R^\mu{}_{\vert\alpha}$ |
| Induced metric: | $g_{\alpha\beta} = \tau^\mu_\alpha \tau_{\beta\mu}$ |
| Outward unit normal: | $n^\mu(u^\alpha)$; $n^2 = -1$, $n \cdot \tau_\alpha = 0$ |
| Coefficients of curvature: | $h_{\alpha\beta} = -n \cdot \tau_{\alpha\vert\beta}$ |
| Area element: | $da = \sqrt{\vert g\vert}\, d^{n-1}u$; $g = \det(g_{\alpha\beta})$ |
| Christoffel symbol: | $\left\{ {\alpha \atop \beta\gamma} \right\} = g^{\alpha\rho}\left[g_{\beta\rho\vert\gamma} + g_{\rho\gamma\vert\beta} - g_{\beta\gamma\vert\rho}\right]$ |
| Covariant derivative: | $V^\alpha{}_{\Vert\beta} = V^\alpha{}_{\vert\beta} + \left\{ {\alpha \atop \beta\gamma} \right\} V^\gamma$ |
| Covariant spinor derivative: | $D_\alpha = \partial_\alpha + \frac{i}{2}\sigma^{\mu\nu} n_\mu n_{\nu\vert\alpha}$ |

(Notation: $A_{\vert\alpha} \equiv \dfrac{\partial A}{\partial u^\alpha}$ for any quantity A.)

The induced metric tensor $g_{\alpha\beta}$ and its inverse $g^{\alpha\beta}$ are used, in the usual way, to transform between the covariant and contravariant forms of tensors. The tensor $h_{\alpha\beta}$, called the "second fundamental form", describes the local curvature of the surface. At any point, the principal values of $h^{\alpha}{}_{\beta}$ are the reciprocal radii of curvature of the surface. Along a timelike direction, the reciprocal radius of curvature is proportional to the normal acceleration of the corresponding spacial surface in its local rest frame. The quantity k is then

$$k = \frac{1}{2}\,\partial_{\|\mu}(n^{\mu}) = \frac{1}{2}\,(\tau^{\alpha})_{\mu}\,\partial_{\alpha} n^{\mu} = \frac{1}{2}\,h^{\alpha}{}_{\alpha}$$

Thus k is proportional to the mean curvature of the surface at each point.

Using the notation of Table I the bubble equations of motion can be rewritten

(I) $\quad i\not{n}\psi = \psi$

(II) $\quad i\not{D}\psi = 0$

(III) $\quad \delta \int d^{n-1} u \sqrt{|g|}\,[\bar{\psi} i\not{D}\psi - C] = 0$

The Dirac equation (II) has a clear interpretation as that of a free massless fermion confined to a curved surface. The equation of constraint (I) on the Dirac field is consistent with the equation of motion (II) by virtue of the relation $\not{D}\not{n} = -\not{n}\not{D}$.

The equation of motion arising from the variation over $R^{\mu}(u^{\alpha})$ is now straightforward to derive. For

$$R^{\mu}(u^{\alpha}) \to R^{\mu}(u^{\alpha}) + \delta R^{\mu}(u^{\alpha})$$

after using (I) and (II), we have

$$\frac{1}{\sqrt{|g|}}\,\delta\left[\sqrt{|g|}\,(\bar{\psi} i\not{D}\psi - C)\right] = -T^{\alpha\beta}\,\tau_{\beta}\cdot\delta R_{|\alpha}$$

where

$$T^{\alpha\beta} \equiv C g^{\alpha\beta} - \mathrm{Im}\,\bar{\psi}\not{\tau}^{\alpha}\partial^{\beta}\psi$$

$T^{\alpha\beta}$ is the canonical energy-momentum tensor of the bubble.

The corresponding equation of motion is

$$0 = \frac{1}{\sqrt{|g|}}\left(\sqrt{|g|}\;T^{\alpha\beta}\,\tau^{\mu}_{\beta}\right)_{|\alpha} = T^{\alpha\beta}{}_{\|\alpha}\,\tau^{\mu}_{\beta} + h_{\alpha\beta}T^{\alpha\beta}n^{\mu}$$

The tangential component of this equation, $T^{\alpha\beta}{}_{\|\alpha} = 0$ follows from (II). This simply reflects the fact that an infinitesimal tangential variation of $R^{\mu}(u^{\alpha})$ is equivalent to an infinitesimal coordinate transformation—the surface itself is unchanged. The normal component of this equation provides the third equation of motion in local form:

$$\text{(III)} \qquad h_{\alpha\beta} T^{\alpha\beta} = 0$$

This equation has a simple physical interpretation. The contraction of spacial components of $h_{\alpha\beta}$ and $T^{\alpha\beta}$ gives the net normal force density exerted on the surface due to its stresses. The orthogonal timelike component gives the rate of change of normal momentum density. Equation (III) is nothing more than Newton's Second Law on a relativistic hypersurface under stress.

The charge, momentum, and angular momentum of the bubble may be expressed in terms of surface fields. In the original field theory these quantities are spacial integrals of densities which are very sharply peaked at the bubble surface. As in the case of the action, the integral over the normal coordinate, $\xi$, can be performed explicitly, to lowest order in D/R, leaving an expression which involves only surface quantities. The resulting charges agree with those derived canonically from the bubble action principle (14). These are given in Table II.

Table II

Conserved Currents and Charges

Fermion Number:

$$J^{\alpha} = \bar{\psi}\gamma^{\alpha}\psi$$

$$Q = \int d\Sigma_{\alpha} \sqrt{|g|}\, J^{\alpha}$$

Energy-Momentum:

$$p^{\alpha\mu} = T^{\alpha\beta}\tau^{\mu}_{\beta}$$

$$p^{\mu} = \int d\Sigma_{\alpha} \sqrt{|g|}\, p^{\alpha\mu}$$

Lorentz Rotations:

$$\mathcal{M}^{\alpha\mu\nu} = R^{\mu}p^{\alpha\nu} - R^{\nu}p^{\alpha\mu} + \frac{1}{4}\bar{\psi}\left\{\gamma^{\alpha},\sigma^{\mu\nu}\right\}\psi$$

$$M^{\mu\nu} = \int d\Sigma_{\alpha} \sqrt{|g|}\, \mathcal{M}^{\alpha\mu\nu}$$

The theory we have developed is manifestly Lorentz invariant and generally covariant. Mathematically, this is a trivial consequence of the fact that all quantities are represented as tensors under Lorentz transformations and under internal coordinate transformations. We note that the spinor $\psi$ is a spinor only in Minkowski space; it is a scalar with respect to surface coordinate transformations. One immediate consequence of Lorentz invariance is that static solutions, which have zero spacial momentum, correspond to particles of mass equal to their energy.

The conserved currents are tangential to the surface at each point. This is a physically and mathematically sensible result. If a current had a normal component, one would hardly expect that its charge could be conserved on the surface. Mathematically, only a tangential current can be integrated over a spacelike cut to produce a conserved "charge". The condition which insures that the conserved currents are tangential is Eq. (I). This equation of constraint on the Dirac field severely restricts the possible fermionic currents that can be constructed. Essentially, we have a two-component fermion. From (I) and the relation $\left\{ i\not{n}, \not{t}^{\alpha} \right\} = 0$, we have:

$$\bar{\psi}\not{t}^{\alpha_1} \ldots \not{t}^{\alpha_n}\psi = 0 \qquad \text{if n is even}$$

$$\bar{\psi}\not{t}^{\alpha_1} \ldots \not{t}^{\alpha_n}\gamma_5\psi = 0 \qquad \text{if n is odd}$$

Thus (I) guarantees that the usual fermion current agrees with the Noether current derived above:

$$\bar{\psi}\gamma^{\mu}\psi = \bar{\psi}\left(\not{t}^{\alpha}\, \tau^{\mu}_{\alpha} - \not{n}n^{\mu}\right)\psi = J^{\alpha}\tau^{\mu}_{\alpha}$$

In contrast, the "axial current" $\bar{\psi}\gamma^{\mu}\gamma_5\psi$ is purely normal:

$$\bar{\psi}\gamma^{\mu}\gamma_5\psi = \bar{\psi}(-\not{n}n^{\mu})\,\gamma_5\psi = \bar{\psi}(-i\gamma_5)\psi n^{\mu}$$

This axial current cannot be "conserved" in any sense in this theory, nor can a Lorentz and coordinate invariant integral over it even be defined. Every current constructed from the Dirac field can be expressed in terms of the vector and pseudoscalar currents. We note that the scalar current vanishes, so that this lagrangian cannot be modified in such a way as to give surface quarks an effective mass.

The generalization of the bubble equations to the case of several independent quark species is completely straightforward. Each quark field appears in the action separately,

$$S = \int du\, \sqrt{|g|}\left(\sum_{\alpha} \bar{\psi}_a\, i\not{D}\psi_a - C\right)$$

Therefore, each quark field obeys the equations of motion (I) and (II), while the fermion contribution to the stress tensor in (III) is the sum over all species.

In the bubble model of hadrons proposed by BCDWY, strong color gauge interactions are introduced which serve to unbind all states which are not singlets under $SU(3)_{color}$. The energy of color singlet states remains unmodified, at least at the semiclassical level. Thus, the BCDWY scheme is equivalent, for our purposes, to a bubble theory of three independent quarks of different colors with the additional selection rule that only color singlet bubble states are allowed.

## IV. BUBBLE STATES

In this section and the next, we consider several exact and approximate solutions to the bubble equations (I), (II), and (III). Before examining these solutions in detail, it is important to recognize the relevance such solutions to a model of hadron structure based on the bubble. The semi-classical theory accounts approximately for the quantum nature of the quarks. The bubble surface motion is treated entirely classically. Classical surface motion is inconsistent in principle with quantized Dirac fields. In practice, we will see that the semi-classical theory has a continuous spectrum of surface excitations and that, although the theory is Poincaré invariant, the states of the theory do not transform as irreducible "particle" representations of the Poincaré group.

In the case of a static surface, the geometric formalism introduced in Section III simplifies considerably. By virtue of Eq. (I) we can write the Dirac field in terms of a two component spinor, $\chi$:

$$\psi = \frac{1}{\sqrt{2}} \begin{pmatrix} \chi \\ i\hat{n}\cdot\vec{\sigma}\chi \end{pmatrix} e^{-iEt} \tag{15}$$

where we have used the Dirac representation of the gamma matrices. In terms of $\chi$, the Dirac equation is

(II') $$H\chi = E\chi$$

with Hamiltonian

$$H = k - i\vec{\sigma}\cdot(\hat{n}\wedge\vec{\nabla}_{\parallel})$$

The normalization of $\chi$ is

$$Q = \int du \sqrt{|g|}\, \chi^{+}\chi = 1$$

and the total energy is

$$U = E + CA$$

The requirement that the action (14) be stationary is equivalent to the condition that the total energy, U, be stationary under all variations of the spacial surface:

$$\text{(III')} \qquad \frac{\delta}{\delta \vec{R}\,(u^{\alpha})} U = 0$$

The system of coupled equations (II') and (III') is very difficult to solve exactly or approximately in three dimensions. Before attacking the three-dimensional problem it is instructive to consider the two-dimensional case, where an exact general solution is available.

In two space dimensions, the bubble is a closed curve in the x-y plane (Fig. 3). We can choose the single parameter describing this curve as its length, $\ell$.

$$\vec{R} = \vec{R}(\ell)$$

$$\hat{e} = \frac{d\vec{R}}{d\ell} = \text{unit tangent vector}$$

$$\hat{n} \wedge \hat{e} = \hat{z}$$

The curvature is then

$$k = \frac{1}{2}\hat{e} \cdot \frac{d\hat{n}}{d\ell} = \frac{1}{2}\frac{d\Phi}{d\ell}$$

where $\Phi$ is the angle of the normal with respect to some fixed direction in the plane (Fig. 3).

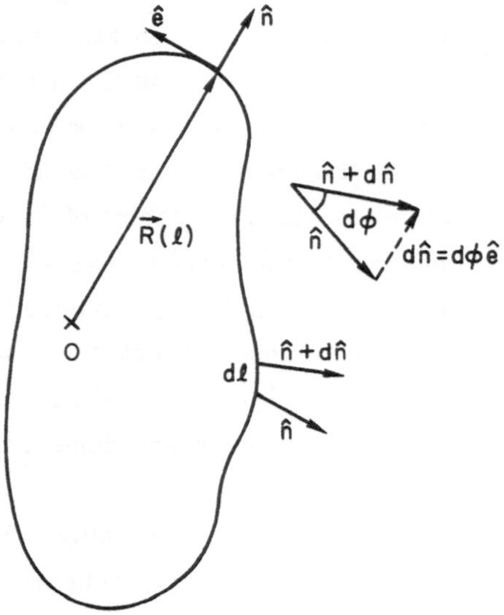

Fig. 3 The two-dimensional bubble.

The Dirac equation is

$$\left[\frac{1}{2}\frac{d\Phi}{d\ell} - i\sigma_3 \frac{d}{d\ell}\right]\chi = E\chi$$

The normalized Dirac wave function can be written

$$\chi = \frac{1}{\sqrt{L}}\, e^{i\sigma_3\left(E\ell - \frac{1}{2}\Phi(\ell)\right)} u$$

where L = total length, u is a fixed unit spinor and E is fixed by the requirement that $\chi$ be single valued: $E = \frac{2\pi m}{L}$ ; m = half-odd integer.

The Dirac energy depends only on the perimeter of the bubble, L, not on its shape. There are paired positive and negative energy levels of the same magnitude. There is no zero energy mode. These results can be readily understood geometrically. A one-dimensional manifold has no intrinsic curvature; from the point of view of a quark trapped on a curve, the geometry in the neighborhood of any point is equivalent to the geometry in the neighborhood of any other point. This leads to a "translation" invariance along the curve. For spinors, this translation is realized by parallel transport, under which the spinor changes only in phase. The Dirac Hamiltonian is just the generator of such translations. Because the quark has spin 1/2, transport around a closed path induces a phase factor -1, which must be compensated by the factor $e^{i\sigma_3 EL}$. Hence, the energy cannot vanish.

We interpret negative energy quark states as positive energy antiquarks. The total bubble energy is, then

$$U = \frac{2\pi\,|m|}{L} + CL$$

Minimizing over L, we have

$$L = \left[\frac{2\pi\,|m|}{C}\right]^{1/2}$$

$$U = (8\pi C)^{1/2}\,|m|^{1/2}$$

It is straightforward to check that, if L is chosen to minimize U as above, Eq. (III') is satisfied at each point on the bubble surface.

The two-dimensional bubble is then extremely soft. Static bubble states occur only with perimeters fixed by the Dirac quantum number m; but bubbles of all shapes with this perimeter are degenerate classically. It is not to be expected that a fully quantized theory will have such an infinite degeneracy. The reflection of the bubble's softness there lies in the large quantum fluctuations of the surface. We shall find that the three-dimensional bubble is also soft, but not so soft that all shapes are degenerate.

We note that there is one conserved quantity which does depend on the bubble shape. This is the angular momentum, $J_3 = M^{12}$. We have

$$J_3 = \frac{E}{L} \langle\sigma_3\rangle \oint d\ell\,[\vec{R} \wedge \hat{e}]_3 = \frac{E}{L} \langle\sigma_3\rangle A$$

where A is the total area of the bubble, and, of course, depends on its shape. Using the expression for E, we can rewrite this result:

$$J_3 = |m|\,\langle\sigma_3\rangle \left[\frac{A}{\pi\left(\frac{L}{2\pi}\right)^2}\right]$$

or

$$J_3 = (8\pi C)^{-1} U^2 \langle\sigma_3\rangle \left[\frac{A}{\pi\left(\frac{L}{2\pi}\right)^2}\right]$$

The ratio $A/\left[\pi\left(\frac{L}{2\pi}\right)^2\right]$ is the ratio of the area of the bubble to the maximum area it could have, given perimeter L. The curve of maximum area with fixed perimeter is unique—a circle. Thus, the maximum possible angular momentum of a state of energy U is

$$J_{3,\max}(U^2) = \frac{1}{8\pi C} U^2$$

Thus the leading Regge trajectory of the two dimensional model is nondegenerate and linear in $(\text{mass})^2$ with slope $(8\pi C)^{-1}$.

Unfortunately, the static bubble equations in three dimensions are not so easily solved. The only known exact solution is a spherically symmetric one corresponding to the approximate solution of the field equations found by BCDWY. It is simply a very difficult technical problem to simultaneously solve the Dirac equation and satisfy the condition that the total energy be minimal under all local variations of the surface. In principle, however, one can find all solutions to the static equations as follows: (1) Solve the Dirac equation exactly on a general closed spacial surface, $\vec{R}(u^a)$. Because the surface is compact, the Dirac spectrum is discrete and the energy levels can be labelled by two discrete parameters, $m_1$, $m_2$ such that the Dirac energy is a continuous functional of the surface variables: $E_{m_1 m_2}\left[\vec{R}(u^a)\right]$. (2) Choose which levels are to be occupied by quarks or antiquarks. (3) Minimize the total energy functional,

$$U\left[\vec{R}(u^a)\right] = CA\left[\vec{R}(u^a)\right] + \sum_{\substack{\text{occupied} \\ \text{levels}}} E_{m_1 m_2}\left[\vec{R}(u^a)\right]$$

in the space of functions $\vec{R}(u^a)$.

Such a procedure is much too difficult to be carried out in practice. It suggests, however, a practical scheme for finding the energy levels approximately. Namely, we attempt to carry out the above procedure, not on a general surface, but over a class of surfaces sufficiently limited that the Dirac equation is tractable. We choose a form for the bubble surface that depends on several real parameters, solve for the Dirac energy as a function of these parameters, then minimize the total energy over the parameters that define the surface. Because the total energy functional is positive definite, such a variational estimate of the energy is an upper bound on the energy of the lowest bubble state with the assumed Dirac quantum numbers $m_1, m_2$.

We begin by considering the simplest possible trial surface—a sphere. Let the sphere have radius R and be coordinated by the usual polar angles $\theta$, $\phi$. The two component Dirac Hamiltonian is

$$H = \frac{1}{R}(1 + \vec{L}\cdot\vec{\sigma})$$

Its normalized eigenfunctions and eigenvalues are

$$\chi = \frac{1}{R}\phi^{\ell}_{jm}(\theta, \phi)$$

where $\phi^{\ell}_{jm}$ is the Pauli wave function of spin j, m and orbital angular momentum $\ell$.

$$E = \begin{cases} \dfrac{j+\frac{1}{2}}{R} & \text{if } j = \ell + \frac{1}{2} \\ -\dfrac{j+\frac{1}{2}}{R} & \text{if } j = \ell - \frac{1}{2} \end{cases}$$

We interpret states with $j = \ell + 1/2$ as quarks, those with $j = \ell - 1/2$ as antiquarks. The total energy is:

$$U = \frac{j + \frac{1}{2}}{R} + C\, 4\pi R^2$$

Minimizing over the parameter R, we have

$$R = (8\pi C)^{-1/3}\left(j + \frac{1}{2}\right)^{1/3} \equiv R_0\left(j + \frac{1}{2}\right)^{1/3}$$

$$U = \frac{3}{2R_0}\left(j + \frac{1}{2}\right)^{2/3} \tag{16}$$

This gives the best approximation to the energy of single quark states with the quantum numbers (j, m) over spherical surfaces.

The local equation for the minimization of the total energy is

$$0 = -h_{\alpha\beta}T^{\alpha\beta} = h_{ab}T^{ab} \quad F$$

$$F = \text{outward normal force density}$$

For the spherical quark state (j, m)

$$F = -\frac{2C}{R} + \frac{E}{R^3} |\phi^{\ell}_{jm}|^2 \tag{17}$$

This vanishes locally only if $j = 1/2$ so that $|\phi^{\ell}_{\frac{1}{2}m}|^2 = \frac{1}{4\pi}$ is independent of $\theta$, $\phi$. For $j = 1/2$, the solution obtained by varying over spherical trial surfaces is exact. In the bubble with $j = 1/2$, the net surface tension vanishes locally. Physically, this reflects the exact balance of the uniform surface tension C and the uniform fermi pressure due to the quark field.

In the BCDWY model, the ground state mesons and baryons are composed of color singlet $q\bar{q}$ or qqq states with all quarks having $j = 1/2$. The corresponding multiplets may, of course, be classified as a $\underline{35}$ and $\underline{56}$ under SU(6). The non-trivial (i.e., non-SU(6)) predictions of this model for the ground state are summarized in Table III.

Table III

Ground State Mesons and Baryons

| | |
|---|---|
| (a) | $M_{56} = 1151\text{ MeV} \rightarrow C = 51\text{ MeV/fm}^2$ |
| (b) | $R_{56} = .77\text{ fm}$ |
| (c) | $\mu_p = 3\left[\frac{e}{2M_{56}}\right]$ |
| (d) | $\left(\frac{M_{35}}{M_{56}}\right) = \left(\frac{2}{3}\right)^{2/3}$ |
| (e) | $\frac{\mu^*_{35}}{\mu^*_{56}} = \left(\frac{2}{3}\right)^{1/3}$ , |
| | e.g., $\Gamma_{\omega\to\pi\gamma}\Big|_{\text{bubble}} = \left(\frac{2}{3}\right)^{2/3} \Gamma_{\omega\to\pi\gamma}\Big|_{\text{Becchi Morpurgo}} = 890\text{ keV}$ |

For $j > 1/2$, the surface tension and fermi pressure balance only on the average; there is a tension induced normal force that will tend to distort the surface from sphericity. From (17), we see that this force tends to push the surface out where the quark density is high, and allows the surface to collapse where the quark density is low (Fig. 4). We have made variational estimates of the energies of the $m = 3/2$, $m = 5/2$ levels on oblate spheroids and toruses. It has been found that, though the distortion of the states from sphericity is very large, the shift in the energy estimated on such surfaces differs from the spherical estimate by less than 10% for single quark bubbles and by about 1% for 3 quark bubbles in which only one quark is excited.

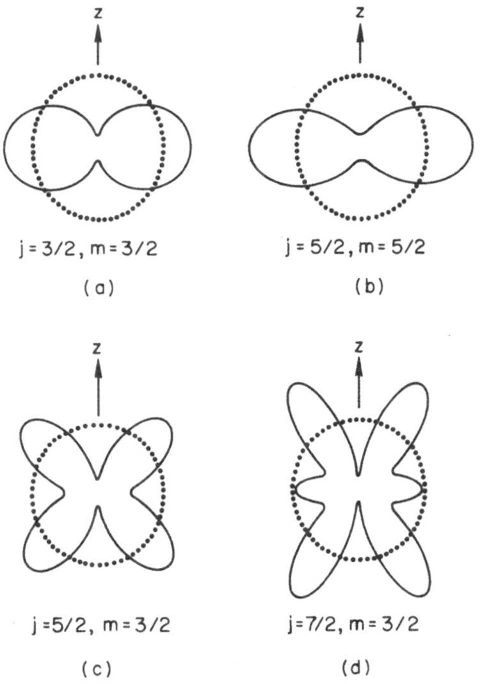


Fig. 4 The normal force density on a spherical bubble for various single quark states. The dotted line is the zero of force.

If we then take seriously the energy estimate of the first, negative parity, baryon multiplet on the sphere, we find that this multiplet consists of the 256 states which can be formed from 2 quarks with $j = 1/2$ and one quark with $j = 3/2$. In our SU(3) symmetric theory, all these states have mass $\left(\frac{4}{3}\right)^{2/3} M_{\underline{56}}$. Relative to the $\underline{70}$, L=1 of SU(6), we have an extra $\underline{10}$, 5/2 and are missing an $\underline{8}$, 1/2 and a $\underline{1}$, 1/2. (See Table IV.)

The MIT bag model, with color gauge interactions and SU(3) breaking introduced via the strange quark mass, has been used in a much more detailed fit to ground state masses.[13)] In this model, we are unable to introduce quark masses by hand so that SU(3) breaking must be introduced though further quark-quark interactions in the bubble.

So far, we have not discussed either semi-classical states in which the bubble surface is nontrivially time dependent or the effects of quantum fluctuations of the surface on the spectrum. It is quite difficult to find nonstatic solutions to the bubble equations (I), (II), and (III). No general prescription for quantizing the surface motion exists.

Table IV

The States of the 70, L=1 of SU(6) Compared to the Bubble[a)]

| SU(3) | $J^P$ | Observed States | | | Bubble Model |
|---|---|---|---|---|---|
| 10 | $5/2^-$ | | | | extra state |
| 10 | $3/2^-$ | $\Delta(1700)$ | $\Sigma(1580)$ | | |
| 10 | $1/2^-$ | $\Delta(1610)$ | $\Sigma(1740)$ | | |
| 8 | $5/2^-$ | N*(1670) | $\Sigma(1765)$ | $\Lambda(1830)$ | |
| 8 | $3/2^-$ | N*(1710) | $\Sigma(1940)$ | $\Lambda$( -- ) | |
| 8 | $1/2^-$ | N*(1660) | $\Sigma$( -- ) | $\Lambda(1670)$ | one missing |
| 8 | $1/2^-$ | N*(1510) | $\Sigma$( -- ) | $\Lambda$( -- ) | |
| 8 | $3/2^-$ | N*(1520) | $\Sigma(1660)$ | $\Lambda(1690)$ | |
| 1 | $3/2^-$ | | | $\Lambda(1520)$ | |
| 1 | 1/2 | | | $\Lambda(1405)$ | missing |

a) From: R. J. Cashmore, "Resonances: Experimental Review," Proceedings of the Summer Institute on Particle Physics, Vol. I, Stanford Linear Accelerator Center report SLAC-179 (1974).

We hope to shed some light on these problems by studying the one class of nonstatic bubbles in three dimensions for which an exact semi-classical solution has been found. These are spherically symmetric bubbles with a time dependent radius—the "breathing modes." We first exhibit the exact solutions of the semi-classical equations. Then we "quantize" the set of all such modes using the WKB approximation. We will see that the "softness" of the bubble is reflected dynamically in the large size of quantum fluctuations of its surface.

We begin with the semi-classical time dependent equations of motion. Let us assume that there is a solution of these equations whose surface is a sphere of time dependent radius R(t):

$$R^{\mu}(t, \theta, \phi) = (t, R(t)\, \hat{r}(\theta, \phi))$$

Define:

$$R = \frac{dR}{dt} = \tanh \omega(t)$$

The Dirac equation can be solved exactly on such a surface for an "L=0" state. This solution is:

$$\psi = \frac{1}{\sqrt{8\pi R(t)^2 \cosh \omega}} \, e^{-i\int_0^t \frac{dt}{\cosh \omega}\left[\frac{1}{R} + \frac{1}{2}\dot{\omega}\right]} \begin{pmatrix} u \\ i\hat{r}\cdot\sigma u \end{pmatrix}$$

where u = any fixed two component spinor. Equation (III) will determine which of the surfaces R(t) are actually allowed dynamical states. Putting the solution for $\psi$ into Eq. (III), we have

$$0 = 1 - R\dot{\omega} - 8\pi C R^3 \left(1 + \frac{1}{2} R\dot{\omega}\right)$$

This equation can be more simply expressed in rescaled variables

$$R_0 \equiv (8\pi C)^{-1/3}$$
$$t = \tau R_0$$
$$R(t) = \rho(\tau) R_0$$

we have:

$$\frac{d\omega}{d\tau} = \frac{1-\rho^3}{\rho\left(1+\frac{1}{2}\rho^3\right)} \;, \qquad \frac{d\rho}{d\tau} \equiv \dot{\rho} = \tanh \omega$$

this can be integrated once to give

$$\epsilon = \frac{1}{\sqrt{1-\dot{\rho}^2}} \left(\frac{2}{3\rho} + \frac{1}{3}\rho^2\right) \tag{18}$$

where $\epsilon$ is a constant. A straightforward integration of the energy density shows that the total energy is

$$U = \epsilon \cdot \frac{3}{2R_0}$$

Thus $\epsilon$ is the total energy of the radial mode measured in units of the static ground state energy.

If $\epsilon = 1$, we recover the static solution, $\rho=1$, $\rho=0$. For $\epsilon<1$, there are no solutions. For each $\epsilon>1$, there exists a unique solution in which $\rho(\tau)$ is periodic, with turning points determined by

$$\epsilon = \frac{2}{3\rho} + \frac{1}{3}\rho^2$$

The equation for $\rho$ is similar to that for a relativistic particle in a scalar potential

$$V(\rho) = \frac{2}{3\rho} + \frac{1}{3}\rho^2$$

shown in Fig. 5.

We note that the total energy is continuous. This is an effect due to the classical treatment of the surface degrees of freedom. In order to get some idea of the level structure of the radial modes, we "quantize" this excitation in the WKB approximation.

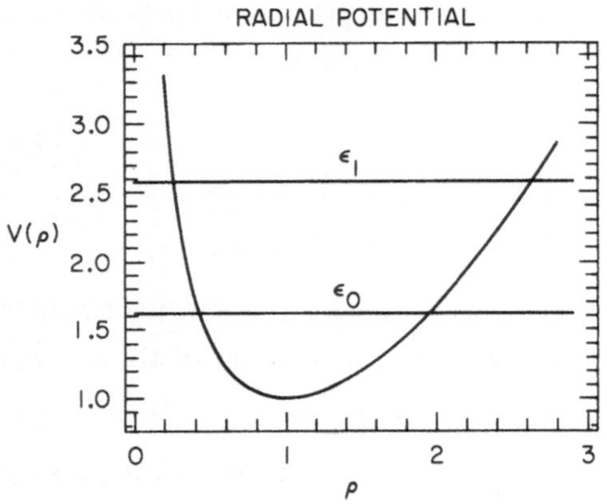


Fig. 5 The scaled radial potential.

We treat the equation for $\rho$ as if it were, indeed, the equation of motion of a relativistic particle in a potential. We take the expression for the total energy (18) to be the Hamiltonian

$$H = \frac{1}{\sqrt{1-\dot{\rho}^2}}\left(\frac{2}{3\rho} + \frac{1}{3}\rho^2\right)$$

The WKB approximation gives the discrete energy levels from the quantization condition

$$2\pi\left(n+\frac{1}{2}\right) = \oint_{\text{orbit}} P d\rho = 2\int_{\rho_{\min}}^{\rho_{\max}} \sqrt{\epsilon_n^2 - \left[\frac{2}{3\rho} + \frac{1}{3}\rho^2\right]^2}$$

This equation can be easily solved numerically. The first few values of $\epsilon_n$ and the corresponding turning points are given in Table V.

Table V

Excitation Energies and Turning Points for the Radial Mode

| n | $\epsilon_n$ | $\rho_{min}$ | $\rho_{max}$ |
|---|---|---|---|
| 0 | 1.615 | .429 | 1.956 |
| 1 | 2.577 | .261 | 2.641 |
| 2 | 3.381 | .198 | 3.081 |

In the lowest state, n=0, we see that the effects of surface zero-point motion are very large. The radius fluctuates by a factor of two about its static value. The energy in the surface excitation is 60% of the static ground state energy. This is a quite dramatic illustration of the softness of of the bubble dynamically and suggests that if fluctuations are properly accounted for, the bubble will be quite smeared out in space.

The n=1 state is the lowest radial excitation of the bubble. Its energy is a factor $\epsilon_1/\epsilon_0 = 1.60$ higher than that of the ground state. It is easy to convince oneself that, in the case where several quarks occupy the lowest level in the bubble, all the energies of the radial mode simply rescale. Thus, the model predicts radial excitations of baryons and mesons with energy 1.6 times higher than the ground state energies.

No radially excited meson candidates have been confirmed experimentally. There is, however, a presumed radial excitation of the nucleon—the Roper resonance—of mass 1470 MeV. We note that 1470/940 = 1.56.

## V. BUBBLES IN 2+1 DIMENSIONS

It is clear from the preceding discussion that any credible theory of quark binding based on the bubble must take into account quantum effects. In particular, the contribution of surface excitations to the bubble mass, the behavior of form factors and structure functions, and the question of bubble interactions can be seriously approached only in a quantum theory of bubbles. The derivation of a quantum theory canonically from the surface action principle is difficult both because the lagrangian is nonpolynomial (in $R^{\mu}$) and describes a "gauge" theory (gauge group of general coordinate transformations in $u^{\alpha}$). No canonical quantum theory has yet been formulated in 3+1 dimensions.

We are fortunate, however, in that a quantum theory of the single bubble can be formulated in 2+1 dimensions. In 2+1 dimensions, the bubble is a closed curve in space which evolves through time—a 1+1 dimensional object. As such, its geometry is relatively simple (e.g., its metric is always conformal to that of flat Minkowski space). In collaboration with S. H. Tye[11], it has been shown that all classical solutions to the field equations (with a two component representation of the Dirac field) can be found explicitly. This complete set of classical solutions can be used as a basis for an expansion in "normal modes", whose amplitudes become the fundamental operators of the quantum theory.

The solution of the field equations is straightforward but tedious. The essential simplification of the geometry arises from the observation that there exists a class of coordinate systems on any solution surface such that

$$R^{\mu}{}_{|0|1} = 0$$
$$(R_{|1})^2 = 0 \qquad (19)$$

These equations resemble very closely the equations describing a Nambu string in 2+1 dimensions in a corresponding coordinate system.

One such coordinate system may be chosen (noncovariantly) and explicit solutions obtained. The classical solutions are characterized by a countable number of parameters[11]: $\{a_n; n=1, 2, \ldots\}$, $\{c_n; n=1, 2, \ldots\}$ and $\{b_m; m=\pm\frac{1}{2}, \pm\frac{3}{2}, \ldots\}$. Physically, $\{a_n\}$ and $\{c_n\}$ describe left- and right-moving surface excitations which propagate with the speed of light. $\{b_m\}$ describes left-moving quarks and antiquarks. In a two-component representation, the condition $i\not{n}\psi=\psi$ removes right-moving fermions. States with $\{b_m \equiv 0\}$ are exactly states of the closed Nambu string in 2+1 dimensions.

A consistent quantum theory is achieved by the introduction of the commutation relations among these modes:

$$\left[a_n, a_m^{\dagger}\right] = n\delta_{n,m}$$
$$\left[c_n, c_m^{\dagger}\right] = n\delta_{n,m}$$
$$\left\{b_k, b_m^{\dagger}\right\} = \delta_{k,m} \quad .$$

The quantum theory has the following general features:

(1) The theory is Poincaré covariant and ghost free.

(2) The normal-ordered mass operator is

$$\mathcal{M}^2 = \mathcal{M}_0^2 + 4\pi C\left[\sum_{n=1}^{\infty}\left(a_n^\dagger a_n + c_n^\dagger c_n\right) + \sum_{m=\frac{1}{2}}^{\infty} m\left(b_m^\dagger b_m + d_m^\dagger d_m\right)\right]$$

$$\left(d_m \equiv b_{-m}^\dagger\right) \quad .$$

where normal ordering is with respect to a "vacuum" annihilated by $\{a_n\}$, $\{c_n\}$, $\{b_m, d_m\}$ and $\mathcal{M}_0^2$ is an undetermined constant.

(3) Physical states satisfy the following condition (reflecting the classical constraint that the surface be closed):

$$\left[\alpha + \sum_{n>0} a_n^\dagger a_n + \sum_{m>0} m\left(b_m^\dagger b_m + d_m^\dagger d_m\right) - \sum_{n>0} c_n^\dagger c_n\right]|\psi\rangle = 0$$

where $\alpha$ is an undetermined normal ordering constant.

As in the string theory, states of the bubble lie on linear Regge trajectories with slope $1/(8\pi C)$. Also like the string in 2+1 dimensions, the parameters $\mathcal{M}_0^2$ and $\alpha$ are not constrained by the requirement that the theory be Poincaré invariant[14)]. Thus, the overall mass scale is left ambiguous. Further, by changing $\alpha$ we may choose to have a theory of bubbles with either even or odd fermion number but not both. $\mathcal{M}_0^2$ and $\alpha_0$ can be chosen to reproduce the semi-classical static energy levels in the quantum theory, with an exponentially increasing degeneracy of states.

Despite the ambiguities in its spectrum, the 2+1 dimensional bubble model is of some interest to us as a guide to the potentialities of the quantum theory in higher dimensions. Particularly interesting is the question of the quantum dynamical properties of the bubble—especially its form factors and structure functions.

At first glance, it appears that the bubble model should give crazy form factors. A thin shell distribution of charge has an oscillating Fourier transform. Further, a quark on a fixed surface has a structure function which does not scale. These observations would be most troubling were it not for the observation that the softness of the bubble must enter bubble dynamics in an essential way. Because the bubble is soft classically, we expect large quantum fluctuations of its surface. This would lead to an effective volume distribution of charge and a smooth form factor. Further, there is no sensible "cavity"[15)] approximation for the bubble—its surface cannot be held fixed in high energy collisions. Indeed, we might expect scaling in the theory only because quarks, in addition to being free to move along the surface, are nearly free to move normal to the surface by dragging it along.

These words, though perhaps comforting, seem to have little change of being supported by quantum calculations in 3+1 dimensions, at least using present techniques. There is some hope that they may be investigated in the 2+1 dimensional theory where the current operator, $J^{\mu}(x)$, has a formal expression in terms of the normal mode operators $\left\{a_n, c_\ell, b_m\right\}$. Unfortunately, $J^{\mu}(x)$ is a highly nonlinear function of the fundamental operators and there are correspondingly severe ordering problems associated with its definition in the quantum the theory[16)]. These problems are currently being investigated.

In this talk, I have tried to give some indication of the properties of states in the bubble model and to discuss some of the theoretical problems involved in such an extended semi-classical model. The discussion of the spectrum was not the kind of detailed confrontation with experimental data which one must have to claim one really has the "sight" model. At this point, I believe such a confrontation is premature insofar as we have not yet straightened out the theoretical question of quantization. I do hope, however, that I have made clear the grounds which we have to hope that there may be a simple and attractive description of hadron dynamics in terms of quarks bound to an extended object.

## REFERENCES

1) W. A. Bardeen, M. S. Chanowitz, S. D. Drell, M. Weinstein, and T.-M. Yan, Phys. Rev. D 11, 1094 (1974).

2) A. Chodos, R. L. Jaffe, K. Johnson, C. B. Thorn, and V. F. Weisskopf, Phys. Rev. D 9, 3471 (1974).

3) Y. Nambu, Lectures at the Copenhagen summer symposium, 1970 (unpublished).

4) String-models with trapped quark wave functions in a sense similar to the bag are discussed by: L. N. Chang, F. Mansouri in Workshop on current problems in high energy physics, Johns Hopkins University, January 1974; and; I. Bars, A. Hanson, YALE-3075-121 (October 1975).

5) H. B. Nielsen, P. Olesen, Nucl. Phys. B61, 45 (1973).

6) R. Giles, Stanford Linear Accelerator Center preprint SLAC-PUB-1682 (submitted for publication).

7) M. Creutz, K. Soh, Phys. Rev. D 12, 443 (1975).

8) Not all variants of bag theories have been derived in field theory. For example, the confined gauge field of the MIT bag.

9) S. D. Drell, Lectures at Erice summer school (1975).

10) S. D. Drell, M. Weinstein, and S. Yankielowicz (to be published).

11) R. Giles, S.-H. H. Tye, Stanford Linear Accelerator Center preprint SLAC-PUB-1687 (submitted for publication).

12) A similar statement could not be made in the quantum theory. It is possible that surface fluctuations always involve local radii of curvature smaller than any finite D. Also, the classical picture of bubble scattering by fusion and fission involves zero radii of curvature.

13) T. DeGrand, R. L. Jaffe, K. Johnson, and J. Kiskis, MIT-CTP-475 (June 1975).

14) In conventional dual string models, $M_0^2$ and $\alpha$ are constrained by commutators $[m^{i-}, m^{j-}]$. Here, there is only one transverse mode so no such commutators of the Lorentz generators arise.

15) C.f., R. L. Jaffe, A. Patrascioiu, Princeton preprint COO-2220-49 (May 1975).

16) Similar problems arise in the case of the MIT bag theory, which has to be quantized in 1+1 space-time dimensions (Ref. 2).

# DESCRIPTIONS OF HADRONIC STRUCTURE*

Y. Nambu

Enrico Fermi Institute of Nuclear Studies
University of Chicago, Chicago, Ill. 60637 USA

I. I would like to devote this lecture to discussing two topics: the evolution of the quark model and the problem of quark confinement. By way of introduction, however, I must stress that after ten years of relative stability, high energy physics is again in a state of flux. At this point, therefore, I had better deal only with generalities. A few significant points characterize the general pattern of the present situation. First there is an evolution of the quark model leading to a proliferation of fundamental constituents. This is of course related to the discovery of new particles. But quantum numbers alone do not seem enough this time, unlike the days of strange particles. New dynamics may also be necessary. An example is the Zweig-Iizuka rule, but asymptotic freedom, quark confinement, etc., must also be included as part of the general problem. One can even speculate that a new post-quantum mechanics is hidden somewhere, although this may be going too far.

The two quantum numbers, color and charm, have been theoretically anticipated for some time. The motivation for color comes from the problem of statistics in the baryon wave function, and that for charm comes, logically speaking, from the weak interactions. There is now more supporting evidence for either concept, but so far the evidence is not direct. This might change any minute. But if color should turn out to be a completely hidden quantum

* This material was prepared for the lectures given at the Workshops in Theoretical Physics, Erice, September, 1975.

number as some theorists desire, the reality of colored quarks will remain a highly theoretical, almost fictitious one. Even if color becomes visible, the quarks could still be confined, retaining their unreal nature. But of course we should not object to such possibilities from our philosophical prejudices. It is an experimental question to be settled with the aid of a theory.

Clearly there is a mismatch between the strong and weak interactions as we know them now. The former exhibits a pattern of SU(3), the latter a pattern of SU(2). It makes our life easy to assign them separate spaces, color and flavor. Each space, especially the flavor, may require further escalation. At the moment a strong and immediate theoretical motivation for it, like those for color and charm, is lacking. But further experimental evidence, such as heavy leptons and more $\psi$'s may force us into that direction. Eventually we will have to worry about a synthesis of all interactions, and how many fundamental particles there are. In this context Gürsey's ideas[1] of using exceptional groups and nonassociative algebras are interesting, albeit highly speculative. They even offer room for post-quantum mechanics. There are also other schemes, such as Pati and Salam's[2], which admit the possibility of integrally charged, non-confined quarks.

Some more details on these escalated schemes. Let me first take the 6-quark model. One theoretical basis for that might be Gürsey's exceptional groups. There are only a finite number, namely five, of such Lie groups, and in that sense it does not lead to an indefinite escalation of symmetries. They all have products of SU(3)'s as subgroups. For example, one of them, $E_6$, contains $[SU(3)]^3$. which may be identified with $SU(3)_L \times SU(3)_R$ of flavor and SU(3) of color.

1. It is interesting to remark that 6 = 2x3 = 3x2. The six flavors may be regarded either as two triplets, each of them satisfying approximate SU(3) symmetry with respect to strong interaction, or as three doublets, for the purpose of weak and electromagnetic interactions. What are the practical uses

of six flavors? First of all, we can build an interesting theory of weak interaction on it, either with V-A currents only, or with both V-A and V+A currents. The latter was recently proposed by Glashow et al.,[3] in order to avoid some embarassing difficulties of the standard charm theory, but it has since gained its own momentum even though other difficulties have cropped up with this scheme.

The V-A theory also has great attraction. Kobayashi and Maskawa[4] remarked in 1973 that if one takes three doublets, each doublet consisting of $Q = 2/3$ and $- 1/3$ members, then one can perform a Cabibbo-like rotation among the three components with the same Q. In general, this will be a SU(3) transformation. In the case of two doublets as in the charm scheme, the corresponding operation is a SU(2) transformation. For the latter, however, any complex phase factors in the transformation can be eliminated by redefining the quark fields, and one is left only with one real Cabibbo angle.

In the case of SU(3), on the other hand, it is not possible to transform away all the phase factors: one is left with three real Eulerian angles and one phase factor. This gives natural room for CP violation as was pointed out by the above mentioned authors.

When it comes to the problem of the $e^+e^-$ annihilation, the 6-quark model certainly will help in increasing the value of R $(= \sigma(e^+e^- \to \text{hadrons}/\sigma(e^+e^- \to \mu^+\mu^-))$, and also can accomodate more narrow states.

2. Next on the integral charge models. The Pati-Salam theory is a logical extention of the Han-Nambu scheme. They include the charmed quark, and regard the four leptons as representing the fourth color, thus recovering the symmetry between color and flavor, or a synthesis of hadrons and leptons. Although they consider both fractional and integral charge possibilities for the hadronic quarks, it seems more natural to take integral charges. It is also natural to regard the color as broken symmetry, so that color becomes visible and the quarks need not be confined.

As was discussed by Pati and Salam, various possibilities of gauging strong, electromagnetic and weak interactions exist. For example, one could conceive of gauging both V and A current for all of them, but this will require a doubling (mirror symmetry) of the fundamental fields to kill the anomalies. The axial vector gluon fields give rise to a spin-spin interaction between quarks. Unfortunately it has the wrong sign, hence no good theoretical motivation. More intriguing and serious is the possibility of violation of baryon number conservation as a result of spontaneous symmetry breaking. However, it will become really interesting only if one can predict the lifetime of a baryon.

3. Since the quarks now seem to have two attributes, color and flavor, one begins to wonder if the quarks might themselves be made up of subquarks, as has occurred to a number of people. Thus one would write symbolically $q \sim f \times c$. Let us suppose in general $q_1 \times \bar{q}_2$ or $q_1 \times q_2$. The conventional theoretical framework dictates that one of them must be a fermion (half-integer spin) and the other a boson (integer spin). I wonder, however, if splitting a quark down the middle into spin 1/4 objects in some sense might not at all be possible. At any rate, the usefulness of the subquark concept may be demonstrated in the example of Gürsey's octonions.[1] In terms of our language, his leptons behave like $\sim q \times \bar{q}$ (integral charge), and his hadronic quarks behave like $\sim q \times q$ (fractional charge), where the components $q$ and $\bar{q}$ have the usual fractional charges.

How can one distinguish between the various schemes? Though the $\psi$ particles and neutrino-induced reactions will obviously be crucial in this respect, at least at the moment no single theoretical scheme can be definitely ruled out or ruled in. This is because, for one thing, our spectroscopic data are still very meager, and besides, we don't quite understand the dynamics such as decay widths and branching ratios. Even if the narrow $\psi$'s are to be identified with charm, color excitations could still show up at higher energies ($\gtrsim$ 4 Gev). Since it will not be profitable to keep on speculating without making specific predictions, I will end this part of the discussion by proposing one extreme color model which M.-Y. Han and I have considered.[5] Assume the quarks to have integral charges and baryon numbers. More specifically, the assignments are:

| | Red | Green | Blue |
|---|---|---|---|
| Q(charge) | (-1,0) | (1,0) | (1,0) |
| Baryon no. | 0 | 0 | 0 |
| Lepton no. | $\pm 1$ | $\mp 1$ | 0 |

There must be an appreciable mixing between q and $\bar{q}$ having the same Q, B and L, which makes the quarks unstable. The quarks with L = $\pm$ 1 are hybrid lepton-quarks which will undergo semileptonic and leptonic decays. The flavor isospin I' is assumed to be a good symmetry. The quark masses will be around 2 Gev.

There will be two-body color excitations of type qq or $\bar{q}\bar{q}$ (color triplet), $q\bar{q}$ (octet) and qq or $\bar{q}\bar{q}$ (sextet) arranged in this order according to the Casimir operator of color SU(3). The triplet qq or $\bar{q}\bar{q}$ is a bound state, but the octet and sextet are unbound resonances.

The $\psi$ and $\psi'$ are identified with the two radial modes of color triplet-antitriplet configuration qq-$\bar{q}\bar{q}$ ($^3P_1$, I=0, C=-1) made of B=0, L = $\pm$ 1 quark pairs. By assumption, the q-$\bar{q}$ mixing is momentum dependent, being large for large momenta (which could be induced, for example, by a short range interaction between quarks.) This allows the wave function to have an appreciable $q\bar{q}$ (singlet and octet) admixture at the origin, which enables it to couple to the photon. But the outer part of the wave function is purer, and all $\gamma$ transitions and hadronic decays are strongly suppressed. The broad resonances above 4 Gev could be interpreted as genuine color octet. Furthermore, the $\mu$-e events reported from SPEAR could be due to pair produced free lepton-quarks. Since one can always form two combinations qq $\pm$ $\bar{q}\bar{q}$, there is a pair of states with C = $\pm$ 1 for a given $J^P$. They must be nearly degenerate since qq and $\bar{q}\bar{q}$ communicate with each other only through the impurity $q\bar{q}$ configuration. But some of the diquark states are exotic, i.e., their $J^{PC}$ cannot be realized by $q\bar{q}$. The $\psi'$, for example, can decay by hadron emission into such exotic levels, which in turn will decay into hadrons. This might account for the "missing" decay modes of the $\psi'$. A possible identification of the observed states according to this model is shown in the accompanying Table and Figure.

II. I will now come to the topic of quark confinement. Its primary motivation derives from the absence of fractionally charged particles, but of course there are also various supporting indications, both experimental and theoretical, which make it an attractive and challenging problem.

As theoretical possibilities for confinement, one may list three categories:

a. Formal arguments like the existence of Dirac monopoles with $g^2/4\pi =$ $= 137/4$ or $137$, and the consequence of non-associative algebras (post-quantum mechanics).

b. Quantum mechanical confinement due to infrared slavery.

c. Confinement picture which works already at the classical level, such as strings and bags.

Common to b) and c) is the use of color gluon gauge fields. Skipping the case a), let me comment on b) and c). The case b) is characterized by unbroken color gauge symmetry which has the elegant feature of combining ultraviolet (asymptotic) freedom and infrared slavery. Unfortunately, the latter aspect is not well understood yet. The behavior of confining forces and the description of hadronic wave functions are not explicit. In this context, the often assumed linear potential ($\propto r$), which has a dimensional constant, and the infrared catastrophe (e.g., Cornwall and Tiktopoulos)[5] are two different mechanisms of confinement. It appears to me that since hadrons have a finite size, the main part of confining mechanism must depend on a dimensional parameter in an essential way, even if there are additional effects which defy classical analogy.

This brings me to case c), which offers a priori various degrees of confinement:

1. Color invisible.
2. Color visible in zero triality sector (broken or unbroken color symmetry).
3. Color visible in all sectors, which means free quarks and no confinement.

In an Abelian model system, the analogy with superconductivity, with the addition of monopoles,[6] offers an attractive mechanism because it is based on spontaneous symmetry breakdown following the long chain of theories, Ginzburg-Landau-BCA-Higgs-Abrikosov-Nielsen-Olesen,[7] and leads to the concept of strings with finite thickness, which combines both strings and bags in a natural way. The former is appropriate to excited hadrons, the latter to ground states. The geometrical configuration of baryons in the string model has been an unsolved question, but recently K. Johnson (private communication) and T. Eguchi[8] argue that excited baryons will take a resonating pattern of q-qq string configurations.

The bold approach of Wilson[9] seems rather different except for his use of stringlike paths. But it may actually be a combination of b) and c). Unless they can carry out their program successfully, there remains the basic problem of formulating a consistent non-Abelian analog of the monopole theory. Even the Abelian theory of Dirac has difficulties with quantum mechanical treatment. The MIT bag theory[10] seems to be essentially equivalent to the picture in which quarks move in a cavity made of an electrically superconducting material, but a precise description of the role of the superconductor is lacking. On the other hand, a semiclassical non-Abelian Dirac string leads to severe and unreasonable constraints on the solution (Eguchi).[11]

I would like now to propose the following formulation based on a conventional field theory. Let us define an octet of antisymmetric tensor $G^a_{\mu\nu}$ (a is the color index which will be suppressed below). It is the analog of Dirac's string field. Then write the defining equation

$$m^2 G_{\mu\nu} = mF_{\mu\nu} + [D_\mu(A)B_\nu - D_\nu(A)B_\mu]^\dagger \tag{1}$$

m is a dimensional parameter which makes dim G ~ 1/L, like that of a potential. $F_{\mu\nu}$ is the usual gauge field derived from a vector potential $A_\mu$. $B_\nu$ is an independent magnetic potential, and $D_\mu(A)$ is a covariant derivative with respect to $A_\mu$. The symbol $^\dagger$ means taking the dual tensor.

Next we relate $B_\mu$ to $G_{\mu\nu}$ by

$$B_\mu = - D_\nu G^\dagger_{\nu\mu} \tag{2}$$

Since the right-hand side defines a covariant magnetic current (note $D_\nu F^\dagger_{\mu\nu} \equiv 0$), this is an analog of the London ansatz, or the field-current identity. Substitution of (2) in Eq. (1) turns the latter into a second order wave equation for $G_{\mu\nu}$, with $F_{\mu\nu}$ as its source.

The third equation to be postulated is

$$D_\mu[F_{\mu\nu} - m\, G_{\mu\nu}] = j_\nu(q) + j_\nu(G) \tag{3}$$

where $j_\nu(q)$ is the "electric" current due to the quarks, and $j_\nu(G)$ is a similar contribution from the G field.

Eqs. (1) - (3) follow from a Lagrangian (with $G_{\mu\nu}$ and $A_\mu$ as independent variables)

$$L = \frac{1}{2}(D_\mu G^\dagger_{\mu\nu})^2 - \frac{1}{4}(mG_{\mu\nu} - F_{\mu\nu})^2 - j_\mu(q)A_\mu \tag{4}$$

which resembles those considered by Kalb and Ramond,[12] by Cremmer and Scherk,[13] and by Kyriakopoulos[14] with various motivations. Suppose m = const ≠ 0. Inspection of Eqs. (1) and (2) shows that the combination mG-F satisfies a source-free equation, hence may be taken = 0, which also means $B_\mu = 0$. But then it leads to $j_\nu(G) = j_\nu(q) = 0$. As a consequence, quarks cannot exist in such a medium. This seems to be related to the Dirac "veto" in his monopole theory, namely the condition that an electric charge cannot touch a string. In our case the magnetic charge is carried by the field $G_{\mu\nu}$. Since a field in the classical sense represents continuous charge distribution, the electric quarks have no place to live due to Dirac's veto.

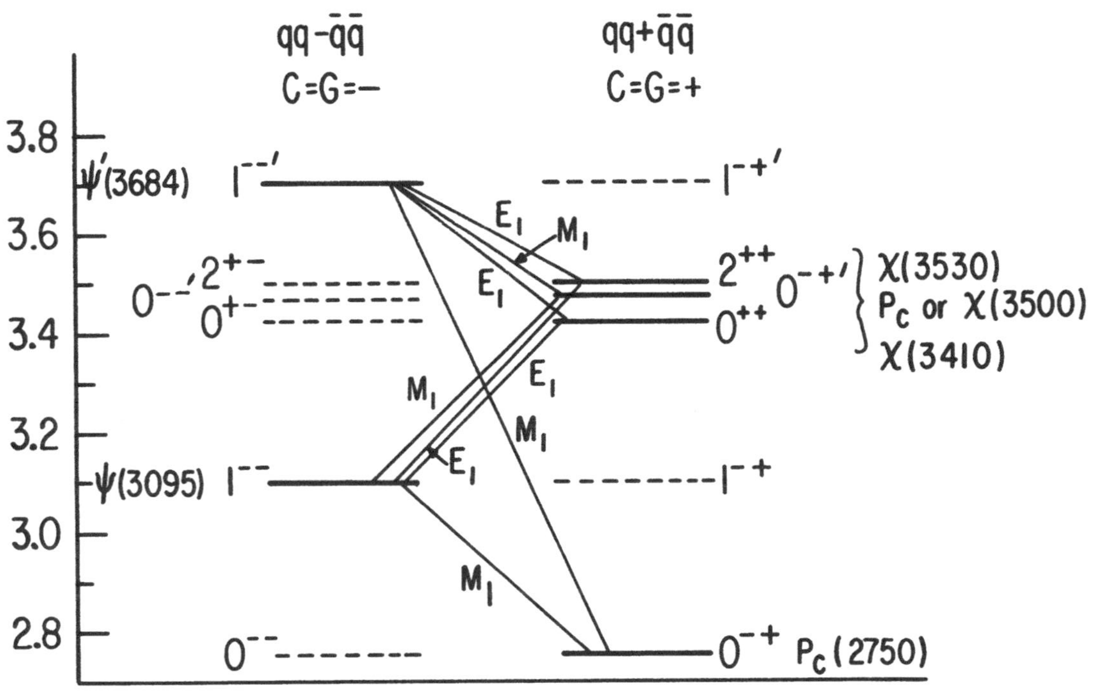


Figure 1. A schematic diagram of $qq \pm \bar{q}\bar{q}$ "chromonium" states achored at $I_f = I_c = 0$ state of $(\bar{3}, 3)$. The "exotic" states are shown in dashed lines. The connecting lines indicate $\gamma$ transitions ($E_1$ and $M_1$ for electric and magnetic dipole respectively.

On the other hand, $G_{\mu\nu}$ and $F_{\mu\nu}$ are decoupled if m=0, and we go back to the usual Yang-Mills theory as far as $F_{\mu\nu}$ is concerned. Now the two cases can be bridged if we replace m by a neutral scalar Goldstone field $\phi$ which develops a nonzero expectation value under normal circumstances. The quarks, however, will create a living space around themselves by making $\langle\phi\rangle = 0$, which will then be the MIT bag, or the Nielsen-Olesen string, as the case may be. The roles of electric and magnetic charges somehow get switched, and the quarks acquire physical electric strings attached to them.[15]

Let us make it a little more precise. Consider the following Lagrangian

$$L = \frac{1}{2}(D_\mu G_{\mu\nu})^2 - \frac{1}{4}\langle\phi\rangle\phi G_{\mu\nu}G_{\mu\nu} + \frac{1}{2}\phi\, G_{\mu\nu}F_{\mu\nu} - \frac{1}{4}F_{\mu\nu}F_{\mu\nu} - i\bar{q}\gamma_\mu D_\mu q - K\bar{q}q \qquad (5)$$

Here $\langle\phi\rangle$ is the vacuum expectation value determined by the nonlinear $\phi^4$ term. Alternatively we can write Eq. (5) as

$$L = \frac{1}{2}(D_\mu H_{\mu\nu})^2 - \frac{1}{4}\langle\phi\rangle\,\phi\, H_{\mu\nu}H_{\mu\nu} - \frac{1}{4}\left(1 - \frac{\phi}{\langle\phi\rangle}\right)F_{\mu\nu}F_{\mu\nu} + \ldots H_{\mu\nu} = G_{\mu\nu} - \frac{\phi}{\langle\phi\rangle}F_{\mu\nu} \qquad (6)$$

Eq. (6) shows that in the asymptotic region ($x \to \infty$, $\phi(x) \to \langle\phi\rangle$) $H_{\mu\nu}$ represents a massive vector field, and the static gauge field energy created by a source blows up like $1/\varepsilon$, where $\varepsilon = 1 - \phi/\langle\phi\rangle$ is the "dielectric constant" of the medium.[16] This means that the static gauge field will be confined, together with its source, to a region where $\phi \neq \langle\phi\rangle$. One can imagine the lines of force squeezed into a bundle or string joining a pair of opposite charges. On the other hand, it is not obvious that the quanta of gauge fields will also be confined. Perhaps a clue to this question can be obtained by defining a new field $\tilde{A}_\mu = \sqrt{\eta}\, A_\mu$, where $\eta = 1 - \phi/\langle\phi\rangle$, in order to renormalize the gauge field kinetic term in Eq. (6). If $\eta \sim e^{-\mu r}$, the field $\tilde{A}_\mu$ develops not only a mass $\mu/2$, but also nonlinear self-interaction terms which are multiplied by inverse powers of $\eta$. It is then likely that the self-interaction will prevent emission of such massive quanta. The physical states will then consist of scalar mesons ($\eta$) and singlet bound states whose constituents are quarks and massive vector mesons (G).

TABLE I

Quantum numbers of some low-lying $qq \pm \overline{qq}$ states with corresponding $q\bar{q}$ analog states

| L | S | qq | $J^{PC}$ of $qq - \overline{qq}$ | $q\bar{q}$ analog | $J^{PC}$ of $qq + \overline{qq}$ | $q\bar{q}$ analog |
|---|---|---|---|---|---|---|
| 0 | 0 | $^1S_0$ | $0^{+-}$ | ---- | $0^{++}$ | $^3P_0$ |
| | | $^3P_0$ | $0^{--}$ | ---- | $0^{-+}$ | $^1S_0$ |
| 1 | 1 | $^3P_1$ | $1^{--}$ | $^3S_1, {}^3D_1$ | $1^{-+}$ | ---- |
| | | $^3P_2$ | $2^{--}$ | $^3D_2$ | $2^{-+}$ | $^1D_2$ |
| 2 | 0 | $^1D_2$ | $2^{+-}$ | ---- | $2^{++}$ | $^3P_2$ |

The Lagrangian (5) looks at least superficially renormalizable. Unfortunately, it is doubtful that this is actually renormalizable. The reason is that the equations of motion lead to certain local constraints among independent variables.

References (Updated in December, 1975)

1. M. Günaydin and F. Gürsey, Phys. Rev. D9, 3387 (1974). F. Gürsey, in Proceedings of the Kyoto Conference on Mathematical Problems in Theoretical Physics, Kyoto, 1975. F. Gürsey, P. Ramond and P. Sikivie, Yale preprint, September, 1975.
2. J. C. Pati and A. Salam, Phys. Rev. D8, 1240 (1973); ibid. D10, 275 (1973).
3. S. Glashow, Phys. Rev. Lett. 35, 69 (1975).
4. M. Kobayashi and T. Maskawa, Prog. Theor. Phys. 49, 282 (1972). S. Pakvasa and H. Sugawara, Univ. of Hawaii Preprint. L. Maiani, Univ. of Rome preprint.
5. Y. Nambu and M.-Y. Han, Univ. of Chicago preprint.
6. Y. Nambu, Phys. Rev. D10, 4262 (1974). G. Parisi, Columbia Univ. preprint CO-2271-29 (1974).
7. H. B. Nielsen and P. Olesen, Nucl. Phys. B61, 45 (1973).
8. T. Eguchi, Univ. of Chicago preprint, EFI 75-50.
9. K. Wilson, Phys. Rev. D10, 2445 (1974). J. Kogut and L. Susskind, Phys. Rev. D11, 395 (1975).
10. A. Chodos et al., Phys. Rev. D9, 3471 (1974).
11. T. Eguchi, Univ. of Chicago preprint, EFI 75-16.
12. M. Kalb and P. Ramond, Phys. Rev. D9, 2273 (1974).
13. E. Cremmer and J. Scherk, Nucl. Phys. B72, 117 (1974).
14. E. Kyriakopoulos, Phys. Rev. 183, 1318 (1969); ibid. D11, 3037 (1975).
15. See Y. Nambu, Univ. of Chicago preprint EFI 75-43.
16. This mechanism was considered by t'Hooft, CERN preprint TH. 1902-CERN, 1974.

# APPENDIX: SUMMARIES OF SEMINARS

An essential part of the activity of the Workshop consisted in the discussion sessions and in short seminars. The editing of the discussions would have delayed the publication of the Proceedings and increased the size of the volume excessively. We therefore decided, regretfully, to omit the discussions and to reproduce the abstracts of only those seminars for which the authors sent us the material in time.

# THE THREE-PARTICLE SPINOR BETHE-SALPETER EQUATION FOR THE BARYONS IN A RELATIVISTIC QUARK MODEL

Manfred Böhm
Physikalisches Institut der
Universität Würzburg
Germany (F.R.)

We extend the relativistic quark model with heavy, tightly bound quarks, which has been worked out for the mesons, to the three-quark problem, the baryons. The dynamics is formulated with help of a three-particle spinor Bethe-Salpeter equation. We assume that only three-particle forces are important and choose for them a harmonic oscillator type form. The spin structure of the interaction is constructed in such a way as to give the SU(6) classification for the baryons, but with amplitudes different from those of Feynman, Kislinger and Ravndal and others. The group-theoretical analysis of the quantum numbers and amplitudes of a relativistic three-particle bound state presented in the foregoing seminar by R. F. Meyer is applied to the BS-equation. It allows us to solve this equation. We obtain - neglecting SU(3) breaking and spin orbit splitting - linear trajectories for the angular momentum excited states. The degeneracy of the higher lying states with respect to their orbital quantum numbers is higher than that of the nonrelativistic model. Finally we discuss in our model some properties of the meson-baryon vertices.

# A RELATIVISTIC QUARK MODEL FOR MESONS AND CP-VIOLATION

D. Flamm
Institut für Theoretische Physik
der Universität Wien

P. Kielanowski
Institute of Theoretical Physics
Warsaw University

A. Morales, R. Nunez-Lagos, and J. Sanchez-Guillen
Departamento de Fisica Nuclear
Universidad de Zaragoza

With a relativistic quark model which is based on the Bethe-Salpeter equation the $K^0$-$\bar{K}^0$ self energy matrix is calculated for the CP violating models of Glashow, Oakes, and Das. For the first two models the kaon mass difference is in reasonable agreement with experiment while for the model of Das it is far too large. The CP violating angles are determined from a least squares fit of the theoretical values of Re $\varepsilon$, $\eta_{+-}$ and $\eta_{oo}$ to the experimental data. While the model of Glashow gives good agreement with experiment, no reasonable fit can be achieved for the models of Oakes and Das. A remarkable feature of the Glashow model is that the CP violating phase angles $\omega$ and $\phi$ turn out to be of the order $10^{-4}$ which is one order of magnitude smaller than originally expected. Furthermore $\omega$ and $\phi$ are nearly equal in magnitude and opposite in sign which is a sufficient condition for the approximate $\Delta I = 1/2$ rule.

(To be published in Physical Review D)

# SPIN AND ORBIT STRUCTURE OF RELATIVISTIC THREE-QUARK AMPLITUDES

R. F. Meyer

Physikalisches Institut der
Universität Bonn
Germany (F.R.)

Motivated by the phenomenological significance of the relativistic quark spin structure of meson and baryon[1] amplitudes, we have analyzed[2] which types of Lorentz and permutation group (S(3)) symmetries are possible for relativistic amplitudes that can be used to describe bound states of three spin 1/2 particles. These depend generally on three Dirac indices and two internal four-momenta.

The transformation behavior in spin space can be classified into six irreps of O(1,3)xS(3), analogously to the five classes of Dirac matrices which are appropriate for the description of the spin structure of a relativistic fermion-antifermion system. The content with respect to O(3)xS(3) is 2·(3/2,Sym) + (3/2,Mxd) + (1/2,Sym) + 3·(1/2,Mxd) + (1/2,Anti), so that the spin structure (3/2,Sym) + (1/2,Mxd) of the nonrelativistic case is considerably enlarged. We have worked out a formalism to treat these classes in dynamical equations with spin-dependent interactions.

The transformation behavior in momentum space is simplified by performing the Wick rotation of the two internal four-momenta.

---

[1] A. B. Henriques, B. H. Kellett, R. G. Moorhouse, Glasgow University, reprint G. U. 38.

[2] M. Böhm, R. F. Meyer, Würzburg University, Preprint R. F. Meyer, BONN-HE-75-14.

Then the orbital part can be expanded in homogeneous harmonic polynomials in eight variables. We have classified these in irreps of SO(4)xS(3) by using an appropriate SU(4) subgroup of SO(8).

Since no (3/2,Anti) occurs in spin space, the construction of an antisymmetric ground state without color requires also in the relativistic scheme a nontrivial orbital part.

# INFRARED STRUCTURE OF LEADING CONFORMAL CONTRIBUTION TO THE ELECTROMAGNETIC VERTEX FUNCTION

P. Menotti

Scuola Normale Superiore, Pisa
Istituto Nazionale di Fisica Nucleare
Sezione di Pisa, Italia

We treat the infrared problem directly related to the asymptotic behavior of the form factor of a particle belonging to the fermion-antifermion channel, with anomalous dimension $\bar{d}$. For $\bar{d} < 2$ we relate it to the off-shell vertex function; for $\bar{d} > 2$ we examine the triangle graph which is considered to be the most infrared divergent. We show that the leading $\gamma_5$-even conformal contribution to the wave function does not give rise to infrared divergences, provided the dimensions of the various fields satisfy the conformal bounds. Thus the ensuing contribution to the asymptotic form factor is $(-q^2)^{1-\bar{d}}$.

# THE ZWEIG RULE, THE OAKES YANG PROBLEM, AND THE $0^+$ NONET

D. G. Sutherland and P. P. Wade-Wright

Department of Natural Philosophy
Glasgow University
Glasgow, Scotland

As explanations of Zweig rule gauge theories have shown promise. As discussed by de Rujula *et al*, a further expectation of gauge theories is that approximately ideal mixing should hold for mesons, $0^-$ apart. Recent analyses by Morgan of $0^+$ mesons indicate a large ($30^o$) deviation from ideal mixing, although the Zweig rule may not be strongly violated. This may indicate a complete failure of narrow resonance ideas or a theory for the Zweig rule other than gauge theories, such as that of Freund and Nambu.

We are investigating whether the ideal mixing and Zweig rule could be true at the quark level, presumably at short distances, but be significantly broken by the coupling to the asymmetric two pseudoscalar channels. This is essentially the old suggestion of Oakes and Yang for SU(3) symmetry. It was shown by Dalitz and Rajasekaran that if the radius of interaction is small this effect can be neglected. Consequent on this and the successes of SU(3), this suggestion has fallen into disuse if not disrepute. However, the case of S-wave final state particles remained an exception for which the Oakes-Yang suggestion may apply. We have investigated this so far in two models, one with heavy quarks in a K-matrix formalism, and the other an R-matrix formalism, with the interior wave function assumed to satisfy ideal mixing and Zweig rule. In neither case is a large enough deviation to satisfy data obtained. One might conclude that this is against models of the type of de Rujula, *et al*, but it may also be argued that our models are not very realistic. We are currently investigating a coupled quark and meson channel model with the quark part of charmonium type.

# MONOPOLE-ANTIMONOPOLE BOUND STATE

R. Ramachandran

Department of Physics
Indian Institute of Technology
Kanpur, India 208016

We relate the monopole-antimonopole bound state mass with the form factor associated with the monopole. We observe that the electromagnetic radius of the monopole $g^2/M$ is much greater than its compton wavelength. The monopole-antimonopole system has relative separation such that it exists mostly within this region. The potential energy of the system is therefore harmonic (linear in a special case) rather than coulombic for the major part.

The size of the monopole is charazterized by the form factor associated with it. The mass parameter in it should be the bound state mass $\mu$. The non relativistic Breit potential obtained from the Born amplitude is, for small separations, linear if the form factor is given by $1/\sqrt{(q^2 + \mu^2)}$ and harmonic if the form factor is either the usual $1/(q^2 + \mu^2)$ or of dipole or multipole nature. We then identify the ground state in such a potential with the meson responsible for the form factor. This results in $M/\mu \simeq 3.2\ g^4$ when the potential is linear, $M/\mu = (3/4)g^2$ for the usual form factor and $M/\mu = (3/32)g^2$ for the case of dipole form factor, where $g^2 = 137$ or 34.3 depending on whether Schwinger's or Dirac's quantisation condition is used. In order to associate such levels with either the newly discovered or older set of vector mesons it is necessary to find a mechanism that would severly inhibit its annihilation which is rather large by virtue of the large value of $g^2$. For further details, see E. Kyriakopoulos and R. Ramachandran, International Centre for Theoretical Physics, Trieste, Italy, Preprint IC/75/150 (to be published in Lettere al Nuovo Cimento).

THE MELOSH TRANSFORMATION IN LOOSELY

BOUND COMPOSITE SYSTEMS

T. Jaroszewicz*

Institute of Nuclear Physics
Krakow, Poland

The transformation is considered between the "classification" and "current" representations of the $SU(2)_W$ group, these two representations being understood as generated by the Foldy-Wouthousen transformed equal-time charges and the null-plane charges respectively.[1] It is argued that, in loosely bound composite systems, where the expansion in powers of 1/c is appropriate, the transformation considered has (up to terms of order 1/c) the same form as the Melosh transformation for free quarks, i.e., that the binding effects are present in order $1/c^2$ only.

This conclusion is arrived at by noting that the transformation properties of a composite system under the group $SU(2)_W$ classification are determined by the system's wave function on the t = 0 plane, in the Wigner spin basis. The transformation properties under $SU(2)_W$ currents, on the other hand, depend on the form of the null-plane ($x_+ \equiv t + z = 0$) wave function, defined in the light-like helicity basis. These two wave functions can be expressed in terms of the covariant (Bethe-Salpeter type) wave function (see, e.g., Ref. 2); in this way the relation between them is found to be given (to order 1/c) by the usual Melosh transformation.

As special example a system of a spin 1/2 particle in an external potential is discussed. It is shown that in this case an

---

*Present address: Max-Planck-Institut für Physik und Astrophysik, D-8000 München 40, Föhringer Ring 6, Fed. Rep. of Germany

_exact_ transformation connecting the "classification" and "current" representations exists but, being dependent on the energy eigenvalues of the system, it cannot be used to diagonalize the Hamiltonian. In the order 1/c, however, this transformation reduces again to the Melosh transformation.

REFERENCES

1. Lectures by J. S. Bell, these Proceedings.

2. T. Jaroszewicz, Lett. Nuovo Cimento _8_, 900 (1973).

PARTICIPANTS

X. Artru
Theoretical Division
CERN
1211 Geneva 23, Switzerland

J. S. Bell
CERN
1211 Geneva 23, Switzerland

R. Benzi
Via Appia Nuova, 154
00183 Rome, Italy

M. Bohm
Department of Physics
University of Würzburg
Rontgenring 8
8700 Würzburg, West Germany

K. C. Bowler
Department of Physics
University of Edinburgh
Mayfield Road
Edinburgh EH9, 3JZ, United Kingdom

F. Buccella
Istituto di Fisica G. Marconi
Piazzale delle Scienze, 5
00185 Rome, Italy

P. Budini
International Centre for Theoretical Physics
Miramare - P. O. Box 586
34100 Trieste, Italy

N. Cabibbo
Istituto di Fisica
Piazzale delle Scienze, 5
00185 Rome, Italy

F. Close
Theoretical Division
CERN
1211 Geneva 23, Switzerland

D. H. Costantinescu
Scuola Normale Superiore
Piazza dei Cavalieri, 6
56100 Pisa, Italy

R. H. Dalitz
Department of Theoretical Physics
12 Parks Road
Oxford, United Kingdom

J. Dunbar
Department of Theoretical Physics
12 Parks Road
Oxford OX1 3PQ, United Kingdom

D. Flamm
Institut Hochenergiephysik
Nikolsdorfergass 18
A-1050 Vienna, Austria

R. Giles
Stanford University
Stanford Linear Acc. Center
P. O. Box 4349
Stanford, California 94305, USA

A. Hey
Physics Department
Southampton University
Southampton, United Kingdom

T. Jaroszewicz
Institut of Nuclear Physics
ul. Radzikowskiego 152
31-342 Krakow, Poland

H. Joos
DESY - Notkestieg 1
Gr. Flottbek
Hamburg 2000, West Germany

L. Maiani
Istituto Superiore Sanità
Viale Regina Elena, 299
00161 Rome, Italy

C. Martinelli
Via Filippo Corridoni, 14
Rome, Italy

P. Menotti
Scuola Normale Superiore
Piazza dei Cavalieri, 6
56100 Pisa, Italy

R. Meyer
Physikalisches Institut der
Universität Bonn
Endenicher Allee 11-13
53 Bonn, West Germany

G. Moorhouse
Department of Theoretical
Physics
University of Glasgow
Glasgow, United Kingdom
and
CERN, The Division
1211 Geneva 23, Switzerland

G. Morpurgo
Istituto di Fisica dell'
Università
Viale Benedetto XV, 5
16132 Genova, Italy

Y. Nambu
Enrico Fermi Institute of
Nuclear Studies
University of Chicago
Chicago, Illinois 60637, USA

L. Oliver
Laboratoire de Physique Theorique
et Hautes Energies - Bat. 2
Universitè Paris Sud
91405 Orsay, France

S. Ono
III. Phys. Inst. der TH-Aachen
51 Jägerst. Aachen, West Germany

R. Ramachandran
International Centre for
Theoretical Physics
Miramare - P. O. Box 586
34100 Trieste, Italy

J. C. Raynal
Lab. Phys. Theorique et
Particules Elementaires
Université Paris Sud - Bat. 211
91 Orsay, France

G. Parisi
Istituto di Fisica G. Marconi
Piazzale delle Scienze, 5
00185 Rome, Italy

H. Ruegg
Department de Physique Theorique
Université de Geneve
Geneva, Switzerland

C. A. Savoy
Deparment de Physique Theorique
Université de Geneve
Geneva, Switzerland

D. G. Sutherland
Department of Natural Philosophy
The University
Glasgow G12 8QQ, United Kingdom

P. Walters
Department of Natural Philosophy
Glasgow G12 8QQ, United Kingdom

INDEX